Alpac

Alpacas as Pets.

Alpaca book for care, costs, behavior, feeding, health, play and exercise.

By

Clive Summerton

Table of Contents

Introduction

As you can guess, this book is all about Alpacas! More specifically, how to keep them as pets in your home or farm. We will delve into every conceivable detail about this animal and its care so that you, the owner, will have a definitive resource at your hands for proper Alpaca husbandry. First things first, Alpacas are *not* Llama's, although the two species are closely related and superficially resemble each other. Indeed, they can be crossbred and produce a (most of the time, more on that later) male hybrid called a *huarizo* or more rarely, a female hybrid called a *misto*. They are different species, analogous to the relationship between horses and donkeys. So what is an Alpaca exactly?

Quick Overview

Alpacas are a member of the family of mammals called *Camelids*. As the name suggests, camels are indeed members of this family of mammals as well.

Members of the *Camelids* family

- Dromedary camels
- Bactrian camels
- Llama
- Guanaco
- Alpaca (of course)
- Vicuña

Alpacas were domesticated thousands of years ago in the Andean highlands of South America, primarily for their fur (called Alpaca fleece). While the native people of the Andes have been enamored with Alpaca fleece for millennia, the rest of the world has taken notice of this beautiful natural resource and Alpacas have become more common throughout the rest of the world for this reason. Alpacas generally have a gentle disposition, but like any other sort of livestock, should not be put in a position where it may feel the need to defend itself. It is far smaller in body size than its more famous cousin, the Llama. Within the pages of this book, you will come to better understand and appreciate the unique background and needs of your alpacas.

Chapter 1: Alpaca's- The Basics

Although you may have experience with other livestock, Alpacas have unique traits that need to be accounted for. The more you know about the background of Alpacas, the better you will come to understand its natural habits and be able to anticipate the needs of your pet more accurately.

Understanding your Alpaca

As stated in the quick overview section- Alpacas are indeed members of a *family* (a taxonomy term used to group like animals) of mammals called *Camelids*. All *Camelids* share a superficially similar appearance characterized by an elongated face and a generally lanky long-legged appearance. The New World members of the *Camelids* lack the distinctive humps so often associated with camels- the nonhump *Camelids* consisting of llamas, guanacos, vicuñas, and alpacas. From a layman's perspective, the four New World *Camelids* look very similar- similar enough in fact that people sometimes fail to realize they are all, in fact, distinct species. Guanacos and vicuñas are strictly wild species, whilst llamas and alpacas are strictly domesticated species.

Natural history and habitat of alpacas

All *Camelids* shared a common ancestor approximately 45 million years ago during the middle Eocene period. The current distribution of the family is, in fact, not at all representative of where the first members of the *Camelids* arose. *Camelids* first appear in the fossil record in the Great Plains region of North America. In fact, *Camelids* were present in North America until the end of the last ice age, some 12,000 years ago during the mass extinction of megafauna that occurred there.

Although there is some debate about why *Camelids,* Ground Sloths, Mammoths, Sabretooth Cats, etc. went extinct in this time frame- most experts can (again, arguably) chalk it up to hunting by the first waves of Native Americans migrating into the region. During the tens of millions of years *Camelids* and other species inhabited the region, there were many climatic and environmental changes, yet they went suddenly extinct when human beings arrived (a pattern that megafauna the world over went through once human beings arrived there). You may ask "Ok, then why did the *Camelids* not go extinct in what would become Latin America?"- frankly, no one knows why! Another (not closely related species) that was apparently obliterated during the end of the last ice age in North America was horses. Again, why they died out in North America yet thrived in Eurasia (only to be reintroduced to the Americas by the Spanish Conquistadors) with a human presence suggests that direct hunting may not be responsible for all the extinctions, but rather an ecological cascade occurred due to the introduction of a super predator (namely, us).

Anyways, *Camelids* were restricted to North America until relatively recently paleontologically. The ancestors to the Bactrian and Dromedary Camel migrated to Eurasia between 3 and 5 million years ago across the Bering land bridge. Meanwhile, the ancestor to the Alpaca (and all other *Camelid* ancestors in the region) migrated to South America roughly 3 million years ago. Eventually, this branch of *Camelids* would become specialized enough to become *laminoids*- a "tribe" (which is a taxonomical designation) under the *Camelid* family. Llamas, alpacas, guanacos, and vicuñas are *laminoids* if you have not guessed it already. Vicuñas and guanacos last common ancestor occurred about 2 million years ago, with the two species diverging ever since.

Natural Habitat

To say that Alpacas have a *natural habitat* is somewhat disingenuous- they are a domesticated species after all. Their natural habitat is wherever human beings will put them. This is not to say that they will naturally thrive wherever a human being will put them. Unsurprisingly, the habitat they will take to most naturally should as closely as possible resemble their immediate ancestor's natural habitat and range- that would be the in the mountains and plateaus of the Andes Mountains in western South America. This is where all *laminoids* can be found- whether domesticated or not. To this very day, this region of the world is where the majority of Alpacas may be found.

The Andes Mountain range stretches from Venezuela all the way through to Chile- although the most pronounced aspects of this range (culturally, geographically and topographically) can be found from Southern Ecuador to Northern Chile. It is the longest continental mountain range in the world at 7000km long (4,300 miles), and is the home to the highest mountain outside of Asia- that would be Mount Aconcagua in Argentina (which stands at 6900 meters/22,800 feet tall). It is also home to the Altiplano, a substantial plateau that stretches across portions of Peru, Bolivia, Chile, and Argentina with an average elevation of 3700 meters (12,300 feet), and is second in size only to the Tibetan Plateau. These facts should give you a clearer picture as to why your Alpaca has certain unique characteristics and needs for proper husbandry.

It is under the above conditions that the vicuña (the wild ancestor to the alpaca) evolved. The vicuña prefers a high alpine habitat ranging in elevation from 3,200 to 5,000 meters (10,500 feet to 16,400 feet) in altitude. The natural range of the vicuña strongly (although not identically) overlaps the Altiplano region of the Andes, with natural wild populations being found in Peru, Bolivia, Argentina and Chile. Their preferred sub habitat within this range tends to be marsh areas- or at the very least, areas with rich, moist soil for the best grazing. Vicuñas are most reliably found within two kilometers (1.2 miles) of a water source. The vicuña is a respected animal in the region, both

7

in Incan and modern times, and despite periods of heavy hunting and habitat encroachment in the past (particularly after the fall of the Incan Empire), it is not currently considered an endangered species, with an estimated 350,000 members alive today.

So when did Vicuñas become Alpacas?
The most immediate answer is about 6000-7000 years before the present, according to the best scientific research available. The *real* answer, however, is the same time as wolves became dogs, boars became pigs and wildfowl became chickens- of course, not all these domestications occurred at the same time, but they did occur under the same circumstances- when human beings came into contact with the wild ancestors of these species and saw utility in their husbandry.

Alpacas and humans
Alpacas and humans have been in close contact for thousands of years in the Andes range, highly valued for the high-quality fiber that can be harvested from them. The fiber is naturally suited for providing insulation in the cold high altitudes of the Altiplano region, and many of the regions world famous brightly colored textiles are in fact derived from Alpaca fiber. Unlike their llama cousins, alpacas, owing to their smaller stature, have never been relied upon as a beast of burden- although they are at times harvested for their meat, just like llamas. The Paleoindians of what would become Latin America probably started herding vicuñas in a similar sort of shepherd lifestyle that emerged with goats and sheep in the Bronze Age Middle East. It has been suggested in academic circles that the Moche culture of Chile may have been the first to domesticate vicuñas- although this is not conclusive.

The earliest reliable records we have of alpacas, and their husbandry, comes from records immediately after the Spanish Conquest of the Inca Empire. The Incan Empire itself was, uniquely among world empires, a preliterate empire, but the Spaniards did take note of the wide variety of Alpaca breeds and phenotypes available in the region. The Spaniards then set about decimating as many alpacas as they could get their hands on to disrupt local economies and promote (the more easily supervisable) European agricultural techniques that would produce goods they would recognize and would have a market back at home for.

Modern history of the alpaca

Unfortunately, after the Spanish conquest of the Incan Empire and surrounding areas, the vast majority of the inhabitants perished. With this human population crash also came a collapse in the indigenous way of life and economy- most of the demand for Alpaca goods collapsed because of these unfortunate circumstances. It has been reported there were a wide variety of Alpacas having been bred and cultivated for millennia with specific characteristics (usually referring to the quality of their fiber) that simply ceased to exist after this period in history. The Spaniards introduced their own forms of domesticated animals, which further competed in the post-conquest colonial economy for Alpaca derived goods. Despite the colonial subjugations of both the peoples and ecology of the Andes, Alpacas persisted in their home range.

In more modern times, Alpacas have been introduced to temperate climates around the world. Unlike many other species (such as maize, turkeys, tomatoes, etc.) subject to the Columbian exchange, alpacas are late beneficiaries, having only been introduced to other parts of the world in relatively recent times. The Spaniards took note of the high quality of alpaca wool in the 18th century, although doubts were expressed that the animals could be successfully conveyed through the tropics, as it meant the alpacas in question would be stuck on a slow-moving ship at high temperature for weeks at a time.

The earliest report of a successful introduction in Europe is in 1772, when "three vicuñas and a guanaco" (probably three alpacas and a llama, considering how difficult it would be to trap and then transport herds of wild animals) were listed as the only survivors of many animals attempted to be shipped that arrived in Cadiz, Spain. One (not specified) died en route to Madrid, another received a cushy home in the menagerie of Charles III at his palace (Casa Del Campo), and the others are simply never mentioned again. Hardly a *resounding* success.

It was the French, under Empress Josephine Bonaparte who attempted a second importation of Alpacas into Europe. She desired a flock for display near Paris, France. In 1802, France and Spain were allied, so chief minister to Spanish King Charles IV, Manuel de

Godoy, arranged for 36 animals (consisting of llamas and alpacas) to be exported from Buenos Aires to Cadiz, Spain. The shipment of nine surviving animals (consisting of llamas, alpacas, and various crossbreeds) arrived finally in 1808 (do take into account how slow bureaucracies are today, add monarchy and sails to understand how such an order could take so long). Unfortunately, the Peninsular War had broken out in the wake of a French invasion in the intervening years, so Franco-Spanish relations were at an all-time low when the animals arrived in Spain. There are conflicting reports as to whether the locals destroyed the animals, or whether they were simply kept in Andalusia, either way, the subject of introducing Alpacas as an additional industry in Europe fell on the backburner in contrast to the Napoleonic Wars raging on the continent at the time.

When the Napoleonic Wars finally died down, the earliest vestiges of the Industrial Revolution would emerge in Europe- primarily in Great Britain. Amongst the first industries to industrialize in Great Britain would be that of textiles- Alpaca fiber first proved to be unwieldy in British hands, but interest persisted in the subject. An entrepreneur by the name of Sir Titus Salt in Bradford, England found a method of weaving Alpaca fiber with cotton that yielded an easier option to machine yarn. This caused an explosion in demand for Alpaca fiber in Britain. Despite this huge surge in demand for Alpaca wool, Alpacas were never introduced en masse to Great Britain, although they started showing up in zoos all over Europe by the mid-19th century, and even Queen Victoria owned two alpacas (one white and one black).

It would not be until truly modern times that alpacas would take off in popularity outside of Latin America. Small private herds had been introduced (and died out again, and reintroduced again in some cases) throughout the United Kingdom, United States, Canada, South Africa, Australia and New Zealand throughout the 19th and 20th centuries, but really only started taking hold by the end of the 20th century. Major importations occurred in 1984 (the United States), 1989 (Australia) for example. In that same year (1989), there were only 150 alpacas in all of the United Kingdom, but a major importation of 300 alpacas in 1995 kicked off the solidification of the alpaca industry in that nation. Perhaps because of the rising numbers and increasing alpaca industry in various nations throughout the world you are interested in owning one as a pet.

Responsible Ownership

Alpacas are classified as livestock almost everywhere they are found, meaning they are not an adequate pet for a suburban or urban lifestyle. Yes, they are diminutive compared to llamas, but they still need an agrarian setting to survive. Alpacas are also herd animals so you would need at least 2 alpacas at a bare minimum to ensure happiness (although the more, the merrier). Although relatively rare, alpaca numbers are climbing, which unfortunately means many people are acquiring the animals without doing proper research into whether they are a suitable pet. Having purchased this book, you are well on your way to responsible alpaca ownership.

Should I adopt or should I find a responsible breeder?
As with any animal, it is always preferable to adopt when possible. Since you are seeking alpaca pet ownership and do not desire a commercial operation, please attempt to contact a rescue organization near you to arrange your alpaca adoption. If you still have your heart on purchasing an alpaca, there are many, many responsible breeders throughout the world who will be more than happy to sell you an alpaca.

Fast Facts

How long will my Alpaca live?
With good care and proper veterinarian oversight, you can expect your alpaca to live 15 to 20 years in captivity. If you decide to own an alpaca, it is a long commitment.

How much will my Alpaca cost?
The quick answer is about as much as any other herd animal that you can classify as livestock. Their needs are not extraordinary, but an alpaca is far from a goldfish or housecat in terms of maintenance. The alpaca itself can range from free (adoption), all the way to $750,000/ £550,000 (although this extraordinary price was for a valuable herd sire at an auction for commercial alpaca production). If you desire to purchase an alpaca, they are usually between $1,000

(£736) and $15,000 (£11,000) for more attractive specimens, although as with any animal, the price will vary depending on quality and your local market.

Enclosure costs

This depends on the size and specific needs of your enclosure. It is hard to give an exact estimate, as your specific needs will vary. At a bare minimum, you will need to provide a chunk of land (generally at least an acre for a few alpacas) and the capability to fence it off. Luckily, alpacas are small as far as livestock goes, so a 4 foot (1.2 meters) fence is sufficient to keep them enclosed. Beyond that, the costs are totally up to you and your environment. If your area has issues with predation (such as the American intermountain west, or rural South Africa), you should consider more robust enclosures to keep predators at bay, especially at night. Obviously, the size, population density, and tenaciousness of said predators will dictate the cost of a robust enclosure (it is far harder to keep an African Lion at bay, than say, an American Coyote). Weather is another factor. If your area suffers from extremes in temperature, you will have to adjust accordingly. Although more robust against cold than heat, you do not want to leave your pets in the howling wind unprotected- at least consider a 3 sided shelter for snowy days. If it gets extremely cold in your area (as in the boreal forest areas of Canada), you may want to consider raising a barn to shield your alpacas from the worst of winter. The converse is true if you live in a warm area- shade, access to water and a regular grooming schedule are essential to keep your alpaca healthy in warm areas.

Food costs

Alpacas do not have extraordinary dietary requirements. In fact, they typically eat less than other grazing farm animals, pound for pound. They eat grass and hay primarily, and at that, they usually only eat about a pound or two of hay or grass a day, depending on their body weight (bigger ones eat more). They will need mineral supplements and fresh water, just like any other livestock animal. Basically, the costs boil down to whatever hay and supplements cost in your area.

Veterinary costs

Alpacas are very hardy animals and do not need an excessive amount of veterinary supervision. Outside of specific individual animal concerns (injuries, odd behaviors, etc.), they will only need their scheduled vaccines and yearly visit. Barring any dramatic happenings, the annual costs should not be beyond a few hundred dollars/pounds.

Chapter 2: Preparing for your Alpaca

Alpacas are livestock first and foremost. You will need an agrarian setting to house them adequately. Luckily, alpacas are generally regarded as the least challenging livestock one can own; they are relatively small in size and do not test the boundaries of their enclosures. In fact, alpacas are ideal livestock to introduce children to, due to their naturally gentle temperament. The most challenging part about keeping alpacas as a pet is that you cannot own just one alpaca- even if you have it herded with other livestock species, it will still need the company of other alpacas for its psychological well-being.

Is an Alpaca the right pet?

You need to be frank about the amount of space you can spare for your alpaca. There are no firm figures for *exactly* how much space you need per alpaca- just take into consideration that they will need space to run and graze to be healthy. Consider the animal's dimensions (about 3 feet/80-100 centimeters at the shoulder in height, and ranging in weight from 100 to 180 pounds (45 kg to 81 kg) depending on the gender and specific animal when spacing out a plot for them. Remember you will also need at least two (although 3 is recommended should one die prematurely) alpacas for their mental wellbeing- although more is merrier. They are natural herd animals and thrive in groups.

Beyond your alpacas, you will have to consider the environment they will be housed in. Extreme climates and predator density will have to be taken into account when designing and building their enclosures and housing. Even if you know for a fact your area has little to no predators, it is a pretty safe assumption that wherever you are, there are stray dogs to consider as well. In a battle between an alpaca and a determined stray dog, the alpaca will lose every time. If you intend to mix your alpacas with other species of livestock, you should be careful to know the other animal's temperaments. Alpacas are generally a live and let live sort of animal and can easily mix with most other animals, although some alpaca owners swear you should not mix donkeys and alpacas, or that the donkeys should be relegated to a perimeter pasture physically separated from the alpacas. They claim donkeys can be aggressive and domineer or even harm the alpacas. The jury is out on this claim, as many alpaca owners have also reported not having any problems keeping donkeys and alpacas in the same space. If you should already own any donkeys, the best course of action is to oversee a limited introduction and watch for any signs of aggression between members of the two species.

Responsibilities
There are some alpaca specific considerations you should keep in mind. These are as follows:

- Keeping more than 1 alpaca at all times without exception
- Shearing your alpaca at least once a year
- Keeping your alpaca's toenails trimmed (every 4 months, on or about)
- Keeping your alpaca's teeth trimmed
- Deworming and providing clostridial vaccinations
- Providing supplemental vitamin D depending on your climate

All the above are the bare minimum health responsibilities for your alpacas that require regular intervention on your (or even better, your veterinarians) part.

Proper Bonding Techniques for your Alpaca

Remember, there is no substitute for time spent with your pet. Obviously, the younger you have acquired your alpaca, the easier it will be to bond with it. There are simple ways to make sure your alpaca's come to recognize you as a friend. Like most livestock, alpaca's wild ancestors are prey animals, so your approach to them and body language should be taken into consideration when approaching. As long as you are not walking in an aggressive fashion, or being otherwise spastic in the vicinity of your alpaca, it should be pretty easy to approach. To seal the deal, approach your alpaca with food in hand. After this, spend time with your alpaca, let it become familiar with your scent and voice. In time, it will come to see you as a friend, and you should have no problem in approaching or handling your pet.

Can I bring my Alpaca indoors?

This may seem like a silly question- alpacas are livestock, they can't come in! However, a cursory look through the Internet will yield a treasure trove of photos and videos of alpacas inside of people's homes. Clearly, this is *not* a long-term solution for your pet alpaca. All we can do is advise people how to go about this most responsibly. Your alpacas are livestock- yes they may be fuzzier and friendlier than say, a cow, but they still need certain considerations. If you feel like having them inside your home for a *brief period of time as a gag under your direct supervision*, then, by all means, go ahead. Just be sure to understand there is a chance of property damage and injury to the alpaca itself. Make sure there is no one or anything that can suddenly startle your alpaca when in confined spaces (children running about, territorial dogs, loud action movies playing on surround sound, a toaster oven ready to throw up toast, etc.) unless you want to stampede livestock through your living quarters. Also please be careful with your alpacas nosing about potential food items (fruits, snacks, houseplants, etc.)- they will try to eat it, and there is a more than slight chance that they will become very ill by consuming something they should not have. Some people like to leave their doors open on particularly nice days, and alpacas may enter on their own accord, so keep this in mind if you are against having livestock trotting through your home. Besides the above tips, it is a pretty solid rule of thumb not to permit livestock in your home.

How do I choose the right Veterinarian or breeder?

Alpacas are livestock- a phrase you will repeatedly encounter throughout any literature involving said animals. This being said, you cannot just take your alpaca in a carry case to any regular veterinarian. As such, they will need to be assessed by a large animal veterinarian (even though alpacas themselves are not terribly large).

Selecting a Veterinarian
As with any livestock, it is almost always preferable to have the veterinarian come to you. Most rural areas have specialist veterinarians that will travel to you and do routine maintenance and minor procedures in the field for your whole herd at once. This is a win-win situation. It makes far more economic sense (in terms of travel expenses and opportunity cost) for your veterinarian to do a whole lot of animals at once in a single day, and it is less stressful for both you and your alpacas to have routine interventions occur at home. Obviously, there may be situations where a particular animal needs specialized care due to illness, injury or pregnancy and your veterinarian will come out just to see that one animal, but let us hope this is as infrequent as possible. You should always seek out a fully licensed and accredited veterinarian; chances are very good your neighbors will know a good one. If that should fail for whatever reason, or you are particularly shy, search on the Internet or call up your local agricultural department to get pointed in the right direction.

Must have's: What to buy

The items needed for proper alpaca husbandry are not extraordinarily different from that of most livestock. Although many of the items on the below list may be borrowed when needed or otherwise outsourced, it is critical that you should have ready access to these resources for the wellbeing of your alpaca herd.

- Animal trailer
- Saltlick/ minerals
- Shearers
- Livestock nail clippers

- Tooth trimmer (more on this later)
- Halter

These are just the accessories you will need, clearly much more infrastructure is needed than the brief list above.

Lone male, lone female, a pair or a herd?

It bears repeating- alpacas cannot be kept as solitary pets. They are herd animals and will suffer tremendously if kept alone. This includes if they are the sole alpaca in a herd of another species of animal- they must be kept in like company! This being said, there are major issues with keeping alpacas in mixed gender herds- we will touch upon this behavioral quirk later. You will always want a minimum of two alpacas (and you will always notice they are in each other's vicinity when you get them)- ideal minimum number is three alpacas in case something unfortunate should happen to one of them, that way you are not sent scrambling to find company for your now alone alpaca. Since alpacas are a herding species, they are also a prey species they will suffer severe stress if alone. They are quite jumpy in a herd setting so a lone alpaca will be significantly stressed out. Remember, in their natural habitats, something is always out to get them, and they feel they have safety in numbers. There is no upper limit to the size of an alpaca herd, other than what you can provide for. The more alpacas, the merrier.

Legal issues

Alpacas are universally considered a livestock species- which comes with a certain amount of privileges and restrictions. The good news is almost every major temperate country has domestic herds of alpacas to buy from. The bad news is that the international trade in alpacas is severely regulated and restricted, as is the case with most livestock species. Because most zoonotic illnesses can jump species, the cross-border transport of any sort of livestock is a bureaucratic and practical nightmare. This is for a good reason. However- a single infected animal can devastate an entire nation's agricultural sector. It is for this reason you should only try to acquire your pet alpacas domestically unless you should want to spend thousands of dollars/pounds for permitting, transporting and quarantining your pets before you ever get them home.

Closer to home, because of the designation as a livestock species, do not try to own an alpaca (although you should always have more than one!) in an inappropriate setting. Common sense should dictate that an average homes backyard is an insufficient setting for livestock- not to mention probably illegal. In recent years there has been a movement to allow chickens, rabbits, potbelly pigs and even miniature horses (the lattermost being designated as a suitable replacement for dogs for the blind) to be raised in more suburban settings, and that is fine- but an alpaca is most definitely not a chicken or a rabbit and will need considerably more resources, attention, and space than a simple coop could provide! Clearly, an urban setting is entirely out of the question unless you would like to make it to the oddball section of your local nightly news (there is always an eccentric animal keeper making it to that segment- don't be that person) and incur fines and other government penalties.

Rule of thumb is always to check your local laws about what livestock would be permitted in your area. Usually, if all your neighbors have livestock, it is a pretty safe bet that agricultural activities are permitted in your area. There are transition zones though, especially in American exurbs, where the law is a little grayer. There was recently a case in a Los Angeles suburb where alpacas were permitted only through omission, as the local statutes permit a limited number of certain species of livestock and "other similar animals." It never hurts to ask your local authorities about what is and what is not permitted in your neighborhood.

Legality in the United States (state by state)

This book will only cover the legality of alpacas in the broadest jurisdictions reasonable to cover. There are simply too many municipalities in the world to cover in fine detail without this book becoming an encyclopedia on the legality of alpacas! Always check with your local municipality regarding whether or not alpacas are specifically permitted (or at least tolerated via omission of any hostile laws). Alpacas are relative newcomers to the barnyard so you may need to do a little explaining to your local authorities, don't be surprised!

Alabama
Alpacas are permitted in the State of Alabama. All *Camelids* (including alpacas) bought from out of state must have a certificate of veterinary inspection *and* an individual identification (either by USDA metal ear tag or tattoo). If the alpaca you are purchasing is older than 6 months old (unless it is a castrated male), it must pass a brucellosis test within thirty days of entering the boundaries of the state.

Alaska
Alpacas are permitted, and I quote, "may be possessed, imported, exported, bought, sold, or traded without a permit from the department but may not be released into the wild." The department referenced is probably the state's department of fish and game. It is always a good rule of thumb to not allow the public to introduce species into the wild, good on you Alaska!

Arizona
Alpacas are permitted, and they simply need to come with a certificate of veterinary inspection from their state of origin (if you are inclined to buy one from out of state).

Arkansas
All *Camelids* (including our beloved alpacas) are permitted in the State of Arkansas. However, if bought out of state, they need a certificate of veterinary inspection within 30 days of entering Arkansas. Animals will also have to be individually identifiable if being transported into the state.

California
Please understand that California is the largest state population wise in the union. With that being said, there is a huge constellation of jurisdictions, some being more famous than others for being restrictive with all sorts of things (including what animals you may and may not possess). At a state level only, you do not need any permits to possess an alpaca. Rest assured, this will vary wildly between municipalities.

Colorado

Alpacas are permitted in Colorado- and participation in alpaca related activities is exempt from civil liability! You would be hard-pressed to be seriously harmed by an alpaca (unless, maybe, you were a small child), but it's good to know that litigation won't be possible if your alpaca should harm someone on your property.

Connecticut

Alpacas are permitted in the State of Connecticut. As always, consult with your local jurisdiction. If you happen to be buying your alpacas from out of state, they will need a negative axillary Tuberculin test within 60 days and a negative Brucellosis test within 30 days at 6 months of age or older.

Delaware

Alpacas are permitted in the State of Delaware. If there are any sort of restrictions on their ownership, they are most likely to be municipal.

Florida

The situation in Florida is a touch more ambiguous, as only camels and llamas are explicitly classified as domestic species (therefore not needing a permit to possess). However, the existence of the Florida Alpaca Breeders Association and a large number of alpaca farms in the state suggests that they are A-OK from a legal vantage point. As always, check municipal laws, as Florida has some notoriously litigious jurisdictions within its boundaries.

Georgia

Alpacas are permitted in the State of Georgia. Only regulations regarding their importation from other states exist. All alpacas entering Georgia must have a USDA ear tag, tattoo, microchip or notarized photograph as a form of identification. If wishing to use microchip as a form of identification, you will have to provide the reader. Oddly, castrated males are exempt from any identification requirements.

Hawaii

Alpacas are permitted in Hawaii. Consider yourself lucky as Hawaii generally has some of the strictest animal regulations in the United

States due to its sensitive ecology. Assuming you are buying your alpacas from out of state (extraordinarily expensive in this case, and not suggested since there are a number of established alpaca farms in the state), you will need the following:

- Requires a Certificate of Veterinary Inspection (CVI) that must be signed by a State, Federal, or accredited veterinarian.
- The CVI must be seven or fewer days old and must state that all requirements have been fulfilled and include the statement: Free of external parasites.
- Required to meet the same entry requirements as cattle, except for anaplasmosis testing.

Idaho
Alpacas are allowed in the State of Idaho, where they are classified as livestock.

Illinois
Alpacas are permitted in Illinois; they are classified as livestock.

Indiana
Alpacas are permitted in Indiana. If bringing your alpaca from out of state, you will need to provide a certificate of veterinary inspection and an animal identification. According to the government of Indiana, the official identification will have to conform to the following standards:

- Official ear tag
- USDA approved Tattoo
- Microchips with ISO 11784/11785
- Digital photographs, printed and notarized sufficient to identify the individual animal

If a health certificate is completed for a movement other than to an exhibition, the CVI must, at a minimum, include a physical description to uniquely identify the individual animal, although the use of an official ID is also permitted as well.

Iowa
A permit is not required to own or purchase an alpaca in Iowa. If you should be buying your alpaca from out of state, you will need a certificate of veterinary inspection that will have both the buyers and

sellers primary address. The certificate should have the species name, description, age and gender of the animal(s).

Kansas
Alpacas are permitted in the state of Kansas. There are laws regarding their importation into the state; per the Kansas Department of Agriculture:

- Need a certificate of veterinary inspection (within thirty days of bringing the animal in to the state
- Individual identification
- Tattoo, microchip, registered ear notch or USDA approved ear tags are all acceptable forms of unique identification for an animal

Kentucky
The State of Kentucky welcomes alpacas with open arms! There are no limitations to their importation, transportation or possession.

Louisiana
Alpacas are permitted in Louisiana. They will be required to undergo tuberculosis and brucellosis, with the legislation affecting them being identical to that of cattle.

Maine
Alpacas are permitted in Maine- there is a bustling alpaca industry in the state. There is only one specific reference to alpacas in all of Maine's legislation, and it is in regards to litter laws caused by pets or livestock- pretty dry stuff really. In title 7 (§3907- definitions) of the statutes of the state, all *Camelids* are referenced to as livestock. Oddly, this appears to not always have been the case, as there was a scandal in 2002 when a hunter shot a llama and did not lose his hunting privileges since it was considered "wildlife in captivity" as opposed to a "domestic animal." This incident prompted the change in the laws of the state, indeed classifying *Camelids* as a domestic species. It is an interesting story. In regards to purchasing them out of state, you will most likely need to contact Maine's Department of Fish and Wildlife in regards to any permitting.

Maryland

Alpacas are permitted in Maryland. There appear to be significant tax benefits to alpaca farming in this state. As for buying or bringing animals to Maryland from out of state, you will need to provide proof of a negative tuberculosis test, amongst other requirements.

Massachusetts

No permits are required to possess alpacas. That being said, you will need to check your zoning in order to own a livestock animal. If coming from out of state, all *Camelids* need valid health certificates.

Michigan

Alpacas are permitted in Michigan. According to the Department of Agriculture and Rural Development, if you should want to bring your alpaca in from another state, New World Camelids (alpacas, llamas, guanaco and vicuna) must be accompanied by the following:

- Official interstate certificate of veterinary inspection (ICVI)
- The ICVI must be properly completed by a veterinarian that is licensed in the animal's state of origin and is accredited by the USDA in the animal's state of origin within 30 days prior to importation.
- All alpaca's must be individual ear tagged, and the individual certification must be recorded on the health certificate

Minnesota

Alpacas are permitted in Minnesota, and, as usual, if coming from another state, must be accompanied by a certificate of veterinary inspection.

Mississippi

Alpacas are permitted in Mississippi. Not a lot of legislation on the books for *Camelids* in Mississippi, so as always, contact your nearest agricultural board for information.

Missouri

Alpacas are permitted in Missouri, but all *Camelids* being brought into Missouri must adhere to the following:

- All alpaca's must have an official Certificate of Veterinary Inspection

- The certificate has to include the following information: the common name of the species, sex, age, weight and coloration
- The alpaca in question must also have an official USDA ear tag.

Montana
Alpacas are permitted in Montana. They are explicitly mentioned in livestock laws multiple times, nothing that should affect a casual pet owner (the laws mostly pertain to free range issues).

Nebraska
Alpacas are permitted in Nebraska. If purchasing out of state for whatever reason, they are subject to general importation requirements in addition to a certificate deeming the animals being free of vesicular stomatitis.

Nevada
Nevada has a freewheeling attitude on the subject of alpacas- no permits of any sort are needed! Even for importation into the state!

New Hampshire
Alpacas are permitted in New Hampshire and are defined as livestock. Like so many other states, if bringing them from out of state, you will need to provide a certificate of veterinary inspection.

New Mexico
Alpacas are permitted in New Mexico. Like Colorado, you have immunity from civil liability when it comes to alpaca related activities! Originally the law was designed to protect those who participate in equine sports from civil law suits, but it appears the law was amended some time ago to include *Camelid* species. Again, if you are bringing alpacas from out of state, you will need to provide a certificate of veterinary inspection.

New York
Hopefully, you are not crazy enough to try and have a pet alpaca in New York City (although, oddly enough, they are not specifically prohibited by the city's laws)- it would be a swift way to end up in the oddities page of the local news. As an alpaca owner, you actually have significant leeway when it comes to owning alpacas; there is

state legislation prohibiting "unreasonable local restrictions" on the ownership and handling of livestock (including alpacas). Try to own alpacas in a rural part of the state, however. As for importing alpacas from elsewhere, well, unfortunately, New York State has some of the most specific legislation in the country- it has laws changing the import requirements depending on what state it is. You will have to research what New York State laws say about bringing alpacas from whatever specific state you wish to bring them in from.

North Carolina
Alpacas are permitted in North Carolina. Alpacas are subject to all livestock requirements, including producing the required paperwork for importing them.

North Dakota
Alpacas are permitted in North Dakota and are considered "domestic animals" (like most livestock). You will need a permit to import an alpaca from another state, as well as a certificate of veterinary inspection, individual identification and negative tuberculosis and brucellosis test results within 30 days of importation.

Ohio
You are permitted to own alpacas in Ohio, and in fact, you are also exempt from civil liability from any "equine" related activities (not that I suggest you ride an alpaca). As usual, if you are bringing in your alpaca from out of state, you will need to demonstrate their good health, with veterinary proof to back it up.

Oklahoma
You are permitted to own alpacas in Oklahoma, they are not specifically mentioned in state statutes, but under guidelines put out by the Oklahoma Department of Agriculture, Food and Forestry, you will need to provide a CVI (again, a certificate of veterinary inspection) and individual ID within 30 days of entry.

Oregon
You are permitted to own alpacas in the state of Oregon. If you happen to own alpacas in a rural area, and they get eaten by wolves, you may be entitled to financial compensation by the state.

Pennsylvania

Alpacas are considered domestic animals by the state of Pennsylvania and are thus permitted. The laws for permitting the importation of an alpaca from another state are a little more stringent than can be comfortably reviewed here (needless to say, it includes exams for tuberculosis and vesicular stomatitis). Please reach out to the agricultural department, if you insist on buying or bringing your alpacas in from another state to review the proper guidelines.

Rhode Island

Rhode Island law views alpacas as domestic animals, as such, they are permitted. Import requirements are similar to other states (CVI, etc.), although apparently, you can bring your alpaca in a relatively small trailer. No idea what circumstance brought about that particular piece of legislation.

South Carolina

South Carolina permits the private ownership of alpacas. If bringing your pet alpaca from out of state, they will need a CVI, official individual identification and testing for tuberculosis and brucellosis (unless under 6 months of age for the latter exam).

South Dakota

South Dakota law views alpacas as livestock, so they are A-OK to keep as pets. As you can imagine, South Dakota is a state that depends on the wellbeing of its agricultural sector, so like almost every other state, you will need to provide an approved CVI along with an official individual ID.

Tennessee

Tennessee is an explicitly pro-alpaca state. Besides being permitted for private ownership, the Tennessee Department of Agriculture actually has an entire section devoted to the wellbeing of the states llamas and alpacas. Even if you don't live in Tennessee, it will do you some good as an alpaca owner (or even potential alpaca owner) to review the materials provided on their page. Based off their alpaca directory, you can surmise that alpacas are a booming cottage industry in the state. If you are bringing your alpacas into Tennessee, the importation requirements are not extraordinary, a CVI and an official individual identification (usually consisting of a USDA valid ear tag or microchip) will suffice.

Texas
Texas, like Florida, is an exotic animal hub, so alpacas are definitely permitted. Texas takes the transport of animals within its states boundaries very seriously. Thus far, it appears to be the only state with rules and regulations about moving livestock (including alpacas, which is classified by the Texas Animal Health Commission as exotic livestock) *within* the state, although this appears to be more concerned with state fairs and such rather than moving from private property to private property. Alpacas being brought into Texas need a CVI but do not need testing for either tuberculosis or brucellosis. Please reach out to the agricultural department if you feel you need any clarification.

Utah
Alpacas are permitted in Utah. Unlike many other states, you will actually need to obtain an animal entry permit to bring your alpaca in from out of state, in addition to a CVI (certificate of veterinary inspection), and negative brucellosis and tuberculosis test results as well.

Vermont
Alpacas are allowed in Vermont. Importation laws seemed to be more focused on regular domesticated animals, although Camelids are mentioned in the legislation. It would be best to err on the side of caution and contact your local authorities as to the procedure to import your alpacas from out of state.

Virginia
Alpacas are permitted in Virginia. They need to be proven free of tuberculosis and brucellosis as well, and the CVI (certificate of veterinary inspection). As livestock, they are simply classified as "other ruminants."

Washington
Alpacas are permitted as pets in Washington State. Importation requirements from other states are more straightforward than many other statutes regarding bringing in alpacas from out of state. They

need to be accompanied by a health certificate (presumably, like many other states, a certificate of veterinary inspection, but it is not specifically worded this way) stating that the alpacas are free from both signs of infection and any other symptoms of communicable diseases.

West Virginia

Alpacas are permitted in West Virginia. They simply need a CVI (certificate of veterinary inspection), and a permanent form of ID to be permitted for importation. They were nearly banned in 2014 due to ham-fisted legislation prohibiting most forms of exotic animals- luckily public outcry caused them to not be on the dangerous animals list.

Wisconsin

Alpacas are classified as livestock in Wisconsin and are thus permitted as pets. You will need to obtain an import permit (in addition to other veterinary documents) to bring in an alpaca from out of state.

Wyoming

Wyoming classifies alpacas as captive wildlife. The law itself states you need no permitting whatsoever to own an alpaca, or even to bring one into the state, but it does strongly suggest that "the Wyoming Livestock Board should be contacted regarding their regulations."

Legality in the United Kingdom

The United Kingdom has a long history of rearing wool-bearing animals (although those in the know with alpacas prefer to call their fur "fleece" to distinguish their final product from that of sheep). Alpacas are no exception to this- there is a rather impressive burgeoning industry of alpaca husbandry emerging throughout the United Kingdom. Despite this, all members of Camelids are specifically mentioned as being classified as a "dangerous wild animal" under the schedule classification in The Dangerous Wild Animals Act of 1976. I am sure you will give someone down at the permitting office a good laugh with your inquiry once you bring up you intend to purchase an alpaca but always go through official channels to make sure your animal stays on the right side of the law.

Legality in Canada

Canada is a large and politically diverse country and with a constellation of differing municipalities. To exhaustively cover the pet laws of every little township, city and village in high resolution would frankly be a futile task. Below is the legal status of alpacas in every province and territory. Canada does have a piece of federal legislation called the Animal Pedigree Act of Canada, which permits the establishment and creation of officially recognized registries. The only official registry for alpacas and llamas in Canada is called the CLAA, the Canadian Llama and Alpaca Association. Although you are not forbidden from purchasing or selling alpacas per se from non-CLAA members, representing the animals as being purebred (whether you are selling it, or if they are selling it to you) from a non-CLAA member is illegal. If you intend to buy "officially" purebred alpacas, it will have to be from a member of the CLAA. You can see the alpaca registry rules here at http://www.claacanada.com/.

Alberta
Alberta has a thriving alpaca industry. Alpacas are permitted in Alberta, and there is an estimated population of 13,644 of them in the province.

British Columbia
Alpacas are permitted in British Columbia, and there is a thriving industry there. There was a bit of a controversy over the subject of weddings taking place on alpaca farms (it seemed the Agricultural Land Reserve threw down the gauntlet to the alpaca farm industry-insisting they had to make more money off of raising alpacas than weddings), but rules have recently been relaxed.

Manitoba
Alpacas are permitted in Manitoba, and there is a thriving industry in this province as well.

New Brunswick
Alpacas are permitted in New Brunswick, although they are somewhat rare.

Newfoundland and Labrador
Alpacas are listed as permitted by the Newfoundland and Labrador Federation of Agriculture.

Nova Scotia
Alpacas are permitted in Nova Scotia.

Ontario
Ontario being amongst the largest provinces in Canada in terms of population has a variety of municipalities. Although permitted on a province-wide level, it is strongly suggested to see if you are permitted to own a small herd of pet alpacas in your particular area. There is a strong reason to doubt they are permitted in the more urban areas of the province.

Prince Edward Island
Alpacas are permitted on Prince Edward Island.

Quebec
Alpacas are permitted in Quebec. Similar to the situation in Ontario, the province has a strong tradition of supporting its agrarian industries but has many urban and suburban areas that may not permit the presence of a small herd of alpacas. Please check with your local municipality to see if livestock is permitted on your property.

Saskatchewan
Saskatchewan does permit alpacas.

Northwest Territories
When researching this book, I had originally anticipated there not being an alpaca presence here. I was wrong- Flat World Alpaca Farm located in Fort Smith, just north of the border with Alberta is a fully functioning alpaca ranch. Alpacas are permitted here, although the climate is a touch harsh for them.

Nunavut

Alpacas are permitted, although the region is noted for not having dedicated animal protection legislation, suggesting they are permitted more by omissions than anything else.

Yukon

Alpacas are permitted in the Yukon.

Legality in Australia/New Zealand

Australia has had its ecology impacted by the introduction of one *Camelid* species- the Dromedary Camel. In fact, Australia is the only place in the world where one will find wild feral herds of the animal. You would think this would have caused a series of draconian laws restricting the presence of other *Camelid* species in the region- it has not. In fact, alpacas are arguably being embraced as pets here more so than anywhere else in the world. New Zealand, likewise, has had its ecology impacted by the introduction of herbivorous vertebrates but has also accepted alpacas as a wonderful domestic species. Again, these two nations have vast agricultural areas as well as noted urban areas so you will need to ask your local municipality if a small herd of alpacas will be legally permitted in your neighborhood.

Australia

New South Wales

Camelids are restricted in New South Wales and will need a license to own and a permit to move. This, however, does not apply for alpacas and llamas, which need no paperwork to move or own- congratulations! The local government does acknowledge that alpacas may be susceptible to Bovine Johne's disease, but there are no regulations for moving them in or out of New South Wales (unlike many, many states in the United States). They do provide a detailed guide as to best practices when moving your alpacas, however to improve biosecurity.

Northern Territory

The Northern Territory does need a wildlife permit from the Parks and Wildlife Commission if you wish to import, keep or export alpacas. You will also need to have a property identification code to keep alpacas on your property, regardless of your property's size or their status as livestock or pets.

Queensland
Alpacas are permitted in Queensland although there is legislation affecting how to keep them. You will need to register your property with the local government (a property identification code will be issued), and your animal(s) will be documented in the National Livestock Identification System. There are also import restrictions, similar to the import restrictions of many American states (vets providing a clean bill of health, permitting, etc.).

South Australia
Alpacas are permitted in South Australia. If you are moving alpacas within South Australia, there is no paperwork required. If moving them from out of state, the alpaca(s) must be accompanied by a National South American Camelid Declaration and Waybill. All properties with alpacas must have a property identification code.

Tasmania
Alpacas are permitted in Tasmania. A "Health Certificate for South American Camelids or Camels Entering Tasmania" must accompany the animals. There are also a variety of smaller laws to comply with.

Victoria
Alpacas are permitted in Victoria, but you must obtain a property identification code to keep them.

Western Australia
Alpacas are permitted in Western Australia, although you will need to obtain a property identification code and individual animal identification.

New Zealand

New Zealand is world famous for its sheep, and thus, for its wool. There is a rapidly growing alpaca industry here, owing to the strong overlap in skill sets and infrastructure needed to harvest sheep's wool and alpacas fleece (as well as the similar climatic conditions alpacas need to thrive). Alpacas are, of course, permitted in New

Zealand. New Zealand has some very detailed (although, not unreasonable) laws regarding alpaca (and llama) husbandry- it's worth taking a look at it to make sure you do not get in trouble for not meeting basic New Zealand guidelines. To read more on the subject, be sure to research the Animal Welfare (Llamas and Alpacas) Code of Welfare 2013.

Chapter 3: Bringing your Alpaca Home

As with any new pet, or really, anyone new in your life, the first impression is crucially important. If something should be drastically wrong, a new animal in a new environment is not the way you would like to find out. If you do your due diligence in advance, your pet should have a painless and seamless transition into its new home. Your alpaca(s) require a very similar sort of habitat as that of any other domestic livestock species, but there are some minor considerations to take into account that is species specific. No two individual animals are alike, why should you expect two completely different species should be the same?

Enclosure selection for your Alpaca

Before you even consider purchasing an Alpaca, you need to make sure the site you have in mind is appropriate for them. Check first with your local municipality- many have very particular codes as to what type of animals you may own and how many of them. Once you have cleared any local legal hurdles, only then can you start to plan for the physical infrastructure of your alpaca paddock.

Luckily, it is almost universally acknowledged that alpacas are far easier on fencing than most other domestic livestock. They do not like to test their physical boundaries, unlike cows and horses. They will scratch an itch with whatever they have on hand so this would include fencing, so it is a bit of a myth that they won't test fencing whatsoever. So, if you only intend to keep alpacas in your enclosure, their fencing needs are moderate at best. More on this later in the book. Far more important is the site you wish to enclose to keep your alpaca herd.

First things first, you will need to assess what kind of climate you live in frankly. Then, compare it to the native range of the alpaca, which was exhaustively detailed for you earlier this book. If you happen to be fortunate enough to live in an environment that nicely matches up with the native home range of the alpaca, then your site selection criteria will be easier. The more it deviates, the more adjustments you will have to make to make sure your alpaca herd is comfortable.

No matter where your enclosure is located, it should have two things to help your alpaca adjust to its environment at its will: shade and protection from the elements. Shade can come in the form of trees, open barns, three-sided shelters, whatever. Protection from the elements should always involve a roof of some sort, that way it will protect your alpaca's sensitive fleece from precipitation.

Finally, be considerate of whatever other animals may be in the area for your alpaca. Remember, alpacas cannot be kept as solitary animals, so if you are just starting out in owning alpacas, make sure to at least buy a pair. While alpacas are kind natured and get along well with other livestock, they are no substitute for the company of another alpaca.

Making sure your Alpaca settles in

Assuming you have recently purchased your alpaca, there are a few things you can do to make sure it adjusts to its new surroundings better. Make sure it has ready access to fresh food and water- this will go a long way towards alleviating your new pet's stress. If at all possible, try to introduce it to the other farm animals gradually- although alpacas are gregarious, they are a prey animal. Too much too quickly can stress your alpaca out. Livestock is less stressful to be around; alpacas can quickly pick up on the fact that these other animals tend to be herbivorous. It is people and dogs in particular that they will need time to adjust to. You as the owner should try to approach your alpaca one on one as much as possible. Unlike other livestock, you can safely approach alpacas all by yourself; they are far too delicate to pose any real threat to human life. About the worst you can experience is being spat at, which, although very gross, is generally not life-threatening.

They do charge from time to time, but I have yet to hear of anyone being seriously injured due to a charging alpaca (it is probably underreported, to be honest- the most serious injury would be to the owner's pride!). As for dogs, never leave dogs unsupervised with your alpacas. This is more for your alpaca's safety than your dog's safety. Alpacas are big (relative to the dog), skittish and fluffy, and very much resemble a prey animal a wolf would hunt. Dogs can easily injure or kill alpacas. Once your alpaca is comfortable around you, then other people, then try to introduce it to dogs, strictly supervised, of course!

Chapter 4: Day to day life with your Alpaca

Your day to day routine with alpacas will closely resemble your routine with other livestock. Generally, your schedule will revolve around seeing that your alpacas are OK first, then going about the more mundane activities to practice proper alpaca husbandry.

Making a schedule

Alpacas are creatures of habit. This is probably best evidenced by their defecation habits- they will always choose to poo in the same spot of the yard. All of them. This is actually pretty convenient; you know where you can and cannot step safely. The downside is something has to be done with that pile. Luckily, like most herbivorous animals, their feces make for excellent compost or garden fertilizer. They do not have a set time of day to defecate; they will just go whenever they feel like it.

In case you are wondering, you can train your alpacas to feed at a certain spot, and they will generally know what time feed time is (it is whenever you set it- but if you keep a schedule, they will adjust to it). It is best not to keep your pets guessing and stick to a routine as much as possible. They are pretty bright, and it is not uncommon to train baby alpacas to feed with their mothers (assuming they are past nursing stage), just to make sure they get their fill. Alpacas can be somewhat greedy and will try to steal each other's food whenever possible, though it is extremely rare for this to escalate into any sort of serious dispute between the alpacas themselves.

What to do if your Alpaca escapes!

An escaped alpaca is an emergency. The world is large and dangerous, and there are many things out there that can hurt your alpaca. Due to its delicate bone structure, even moderate injuries can be life-threatening. They are flighty and prone to panic, so cars may startle it. Also, due to their diminutive size, they make a relatively easy prey animal for domestic dogs to attack and kill. Of course, if you live in an area with serious predatory animal activity (cougars, bears, etc.), this only bodes for the worst for your escapee.

First things first, make sure your alpaca has actually escaped. Panic never does anyone any good. If you have a large herd, take the time to make sure it is not laying down amongst some friends, or hiding out in an odd corner of the barn. If you know for a fact that an alpaca has escaped, call neighbors and local law enforcement. An alpaca on the run is a unique sight, and it should not be long before someone makes mention of it to the authorities. This should help get you in front of your alpaca as soon as possible.

The best preparation for an escaped alpaca occurs before your alpaca escapes. If you have spent a lot of time socializing and adjusting your alpacas to a human presence, this will make a world of difference when the time comes to approach it during its great escape. Ideally, your escaped alpaca will have a holster to hook up to a leash, barring that, hopefully, you have trained it to respond to a sound (clicker, for example) during feeding time. If no training has been done, then your best bet is for you as the owner it knows and trusts to approach the escapee with food. Food makes for a wonderful bribe with just about any animal, and alpacas are no exception to this.

Hopefully, your alpaca is within walking distance, and you can simply walk it home. If the alpaca has traveled some distance, you may need to bring your trailer with you during your search (or reclamation if the alpaca has been located) to minimize the stress on both yourself and your pet alpaca. Just keep calm, and keep your animal calm. Alpacas are sensitive to body language, like most prey animals, and are prone to panic if they feel they are in danger.

Traveling with your Alpaca

No way around it, from time to time, you will need to transport your alpaca. Of course, it is almost always better to have whatever services your alpaca needs come to it to minimize stress on the animal and wear and tear on your equipment. There will be instances though where travel is unavoidable (or even desirable!). It can be a stress-free process with the right forethought and equipment.

Where are you taking your Alpaca?
This may sound like a silly question, but it is the most important one. Consider the distance you are traveling with your alpaca. If it is for hours, make sure you bring food and water. If the weather is hot, water is especially important; the animals travel with a practical coat for all year round! The longer the time, the more likely it is you will need your alpaca to travel with a herd mate to minimize stress. If you are traveling across borders of any sort (whether from state to state or even internationally), please make sure you have researched what the required paperwork is and double check to make sure it is filled out correctly and that you remember to bring it with you. Alpacas are considered livestock by most governments and thus almost always need permitting of one sort or another to travel (unlike a cat or dog). An impounded alpaca is a stressed alpaca- and stress is life-threatening for most ruminants. By the time you get everything sorted out with the authorities, it may be too late. Just do the best you can as an owner to put yourself in your alpaca's shoes and think what you would like to avoid (hunger, stress, loneliness, excessive heat or cold, etc.), and you should be fine.

Routine travel with your Alpaca
Ok, let us be frank. Your alpaca may be small and cute, but it is still livestock. Make no attempt to move it without first buying an animal travel trailer (although there are stories of people moving the babies in cars, this is NOT suggested). Even the very young should get used to traveling in the trailer, this is not a process you want an older animal experiencing for the first time and freaking out. You will find videos and images on the internet of alpacas being transported in SUV's and such like a large dog. What those videos and images fail to show is that alpacas are not housebroken (fun!) and what would happen should you have several hundred pounds of panicked alpaca on the loose in the same compartment you are driving in? Please, buy a trailer, for both the safety of you and your alpaca.

Long distance traveling with your Alpaca
First thing's first- see if you are crossing any sort of border. As mentioned above, paperwork is extremely important when moving livestock across boundaries. Some more draconian places may even need paperwork if moving within the state (province, etc.)! Don't let your alpaca be an impounded alpaca because you claim you did not

know or thought you would get away with it. A stressed alpaca is most likely a dead alpaca. After that, plot out how long of a journey you are taking with your alpaca. A good rule of thumb is to pull over and let your alpacas out for a walk roughly every four hours. For example, if you are moving your alpaca from Central California to the Denver metropolitan area that is about an 18-hour drive nonstop. This means you should pull over at least 4 times to let your alpacas out for about half an hour or so. The longer the journey, the better it probably is to break it up into manageable chunks- no one likes being on the road for an extended period of time, your pet alpaca won't either.

The second thing to take into account is the changing climatic conditions along your route. This is probably more important in the United States and Australia than elsewhere, simply due to the humungous size of these countries and the distinct climatic changes one can encounter during a long road trip. If heading from cold to hot weather, make sure to take any sort of blanket off your alpacas when things start to warm up, and vice a versa when going from hot to cold weather. You may also want to put ice in the alpaca's drinking water if you are traveling for an extended period of time in warm weather.

Always travel with multiple alpacas if you are making a long trip- alpacas are gregarious in nature and need each other's company in order to thrive. This is especially true in any situation that may be perceived as stressful, and travel for most animals is extremely stressful. Like with any other pet, just try to put yourself in your alpaca's shoes. Anything you might find annoying, they almost certainly will too.

Chapter 5: Making a comfortable enclosure for your Alpaca

Your alpaca's enclosure is its home. Home should always be comfortable, whether for yourself or your alpaca. Alpaca's are relatively hardy, as far as it comes to climatic conditions they can handle, but this does not mean you should not take some basic steps to enhance their living conditions. It is true, they are native to the high Andes, and are thus adjusted to extreme cold. It is also true that we as human beings are native to the plains of East Africa and its extreme heat waves. Just because our genetic heritage endows us with certain proclivities does not mean that we will necessarily be comfortable in worst case scenarios. Plus, our ancestors and your pet alpaca's ancestors both enjoyed an advantage that modern day humans and alpacas alike no longer enjoy; complete and utter free reign around their environment. Where ever we are now, we are stuck with and have to adjust our environment to our needs. You can bet no vicuna ever willingly stood out in the middle of a blizzard exposed, just like none of our ancestors took the full brunt of the sun without at least searching for shade first. As humans now, we can go into our shelters and adjust the temperature to whatever we like, let us hope the same consideration is afforded to our pet alpacas.

General guidelines for your Alpaca enclosure

Alpacas are not exceedingly difficult animals to keep in. Basic fencing will do the job, and they will not test fencing as much as a cow or a horse. Believe it or not, the fence has to be relatively robust to keep other animals out, as opposed to keeping your alpacas in. The minimum fence suggested should be coyote proof- this means about 5 ft. (1.5 meters) high of woven wire with 2 inches (5 cm) by 4 inches (10 cm) mesh for the outermost perimeter of your alpaca enclosure.

You may say to yourself- "Well, there are no coyotes where I live!" and you may very well be right. However, take a moment to ask if you have ever seen a stray dog where you live? Unless you happen to be in Antarctica (which is uniquely harsh for both you and your alpaca), the answer is going to be "Of course!" who has not seen a stray dog in their life? Coyotes are pretty wily (hence the famous cartoon), and if this western American guideline keeps them out, it should do pretty well with most dogs as well. Your alpaca will not be able to tell the difference between a predatory coyote and a mischievous young middle sized dog- the bites feel identical from its perspective!

If where you happen to live happens to have more serious predation issues, you will have to scale up accordingly. You may want to consider getting a guard dog, or, a guard llama to keep your herd protected while you are away. Be sure to bond your guard animal of choice from a young age to your alpaca herd, however. What makes them intimidating to potential predators could also make them dangerous to your alpaca herd should they not be socialized adequately. There is also special fencing designed to keep predators out, although this is pricier than normal fencing meant to keep livestock in. You may also consider running a (low voltage) electrically charged wire through your fencing- one zap should scare off most predators. By far the most cost-effective way to reduce predation issues is to keep your alpaca herd in a night enclosure. More on that below.

Before I forget, the rule of thumb is 1 acre (.4 hectares) per 5 alpacas at a maximum to keep them happy!

Where should your Alpaca sleep?

If where you live has no predation issues (and you feel comfortable the stray dog issue in your area is well managed), and the climate is favorable (at least during that evening's overnight hours) you may leave them to pasture overnight. This is not strongly suggested, as you never know when some careless person might drop off a pet mastiff in your area (this is especially true in rural areas), but generally, there is no harm in leaving them outside from time to time.

Your outdoor enclosure will need some sort of lean-to at the bare minimum to keep your alpacas comfortable. Make sure there is enough covered space to provide this minimum level of shelter to each and every one of your alpacas- even if this means building additional leans to sheds. If a freak thunderstorm should come, it is important your alpaca stays as dry as possible. Even high levels of wind could be bad for your pet alpacas, as wind can carry dust and pebbles that can disturb your alpaca- a minimal level of shelter will go a long way towards ensuring your alpaca's comfort.

Your alpacas are very light sleepers. In fact, it is pretty hard to tell a drowsy alpaca from a chilling out alpaca from a sleeping alpaca- they always seem to keep their eyes semi-open, and even when they are fully shut, they wake so easily it is hard to tell if you caught them sleeping. This is because alpacas are pretty much defenseless prey animals- they are very easily disturbed. Try to make sure as many unnecessary stimulants as possible are removed from your alpaca's vicinity (wind chimes, etc.), and any nuisance animals are dealt with accordingly.

A barn is necessary for your alpaca, no matter where you live. Barns protect against predation, rain, snow, high and low temperatures. Remember, just like people, your alpaca will not be in perfect health 100% of the time, so somewhere comfortable and private will help your alpaca from becoming overly stressed and help it recuperate more quickly. Adjust your barn accordingly to your environment. If your area suffers from excessive heat overnight (like, say, Florida) consider air conditioning it- ventilation is extremely important for your alpaca, as they can overheat (have you seen their fleece!?). Make sure it is made of solid materials and can withstand bad storms and the persistent attempts of predators. Some alpaca owners have gone as far as to install motion sensitive lights to the exterior of their barns- this will startle the predator and make it feel exposed. I can strongly suggest this technique, as it will ultimately be less stressful for your alpacas than hearing certain death pawing away at the walls and doors of your barn.

Heating and lighting

Alpacas are cold weather animals, so heating needs are minimal. Make sure your alpaca has access to some sort of shelter to at least escape the wind. If you expect your alpaca is still cold, you may put a blanket on it to keep it warm. You may also improvise a jacket for your alpaca- make sure that it is Polartec on the inside and waterproof Cordura on the outside. Make sure you have several handy in this instance, as alpacas are still animals and will soil their jacket from time to time. Access to a barn is best, of course. This is especially true for expecting or baby alpacas, who are way more vulnerable to weather extremes than their adult counterparts. Make sure you have adequate bedding (the common livestock bedding options like straw and the such is perfectly reasonable), and if you are particularly worried about extremes with cold, maybe a heat lamp.

Heat is the real enemy for your alpaca. The high Andes do not get excessively hot, rarely getting above 64 Fahrenheit (18 C)- this is not the same case everywhere else. Also, alpacas are usually found at relatively high altitude, which makes shedding heat even easier- almost certainly wherever you live is closer to sea level than the alpaca's typical range. The best thing you can do during particularly warm seasons (spring or summer, depending on where you are) is to have your alpacas fleece shorn. This removes most of their capability to retain heat and will do a lot to alleviate heat stress from your alpaca.

You can also hose your alpaca down, especially after your alpaca has been shorn (unless you do not want to harvest fleece for whatever reason, as the water will mess it up). Hose their necks, legs, and belly down, in particular, they seem to enjoy it during hot months. Always make sure your alpacas have access to shade as well, whether that means access to a lean-to or a barn. Fans are also a great, relatively low-cost option to add strategically to the enclosure. If where you live is particularly hot and humid, an air-conditioned barn may not be unreasonable. This is a pricey option, but if this is the only way to keep alpacas comfortably in your climate, it is a must-have.

Accessories

So some accessories you should have on hand to properly raise your alpaca will include the following (the list is hardly exhaustive, and some options are optional, but it is a great place to start):

- Halter
- Anti-fungus dip for said halter
- Protective noseband for the halter
- Lead rope
- ID collar
- Utility feeder
- Hook-over trough (useful for multiple alpacas)
- Salt block holder
- Mesh hay bag
- Automatic waterer
- Water safe bucket
- Brush to clean the bucket
- Bucket holder
- Submersible or floating tank deicer, if necessary
- Collapsible water bucket (especially handy if you need extra water due to it being an excessively hot day)
- Camelid scale (very useful for your veterinarian)
- Large poop scooper (remember, they pick a spot and use it as their latrine, that pile has to move eventually)
- Rake (for the bedding)
- Tree wrap (they like to gnaw on bark, they will eventually kill your tree or trees this way)
- Fly mask
- Spit mask (for your unruly alpacas)
- Travel trailer
- Electrolytes
- Probiotics
- Livestock friendly shampoo
- Equine wound spray
- Triple antibiotic ointment
- Blood stop powder
- Veterinarian wrap
- Disinfection mat (especially useful should you be visiting other farms or ranches regularly, you would not want to track back any nasty critters to your herd)
- Ear mite lotion
- Terramycin ophthalmic ointment
- Livestock appropriate dewormer

- Good shearers
- Alpaca insulated coat (they do make them, but improvising is fun too)
- Coat strap extender

Now, as cute as it may be, please do not attempt to pack your alpaca. Llamas can handle holding packs because they are larger and stronger and frankly have a better body plan for packing than alpacas. Alpacas have never been bred as pack animals, and you will not be the first one to pull it off successfully. Don't even think about it!

Housing for your herd

Alpacas are herd animals, they need to be housed as a group with a few exceptions. Clearly, animals in labor will need some privacy, as will the very young initially after birth (they should be in their mother's presence, however). Aside from this, they must spend as much time around each other as humanly (well, rather, alpaca-ly) possible. They are gregarious and need each other's company. So to boil it down, alpacas should have an outdoor enclosure, as well as a solid barn to retreat to should the weather become inclement. It must be at least minimally weather appropriate as well as predator resistant in proportion to what can be reasonably expected in your area (no point in spending a lot of money trying to thwart an escaped circus tiger that may never come). Access to clean food and water is a must, as with any other pet. Finally, you must adjust your alpaca's housing to the climate. If you are prone to harsh winters, you must have an escape for your alpaca from the driving sleet and snow. If your area suffers from excessive heat, you must provide cooling options. See, alpacas are truly some of the lowest maintenance livestock around.

Chapter 6: Alpaca behavior and husbandry

The old saying is an ounce of prevention is worth a ton of cure. Well, preventing misunderstandings with your pet alpaca will spare you much grief. Probably the most difficult thing to do with animals, at least, when it comes to trying to understand them, is to not anthropomorphize them. Your alpaca ancestors evolved under very specific conditions, and the alpaca itself was bred for very specific purposes. This genetic endowment, far and away, will influence your alpaca's behavior far more than its environment. Although they have charming, doughy eyes, understand that your pet is still an animal and has its own prerogatives.

How Alpacas are in the wild

How are alpacas in the wild? Trick question! There are no truly wild alpaca's (all though I am sure a few may have gone feral from time to time- and have probably not fared well). For years, it was assumed the alpaca, and the llama was descended from a recent common ancestor- the guanaco. While it is true, that llamas and alpacas share a common ancestor, it is way further back than the guanaco, which recent genetic research has shown to be the ancestor to the llama exclusively.

The diminutive vicuña has been conclusively demonstrated via modern genetic techniques to be the ancestor of the alpaca. They have extraordinary hearing and are thus easily disturbed. Due to being rather defenseless, they are extremely shy and easily startled. About the only advantage they have when it comes to facing off with predators is sheer numbers- the odds of being picked off drop dramatically the larger the herd you are with. Does all this sound familiar to you yet, in regards to your pet alpaca?

In the wild, they browse for short, dense grasses and are huge fans of salt however they can get it. They tend to lick mineral outcrops for their salt intake, but they have been observed drinking out of saltwater pools in the Altiplano. Vicuñas tend to live in small herds, usually with a lead male and 5 to 15 females. The groups tend to be somewhat territorial, with the average territory being about 18 km

(6.9 miles) sq. Smaller territories are possible, assuming food sources are richer and more densely packed.

There is a defined breeding season for vicuñas, usually occurring in the months of March through April (which is unlike alpacas, which may breed at any time of year). Fawn become independent at one to one and half years of age, and young males will travel the countryside in bachelor groups looking for "sororities" of independent young females to join. Thus new herds are consistently forming while minimizing the risks of inbreeding.

Alpaca communications

Alpacas can be quite vocal. Indeed, they probably make a wider range of sounds than most other barnyard animals, but I would not go so far as to say they are noisier. They make roughly nine different types of noises, and I will do as best as I can to describe them into words.

- **Clucking**- Sounds somewhat like a chicken clucking. They tend to do this in two instances- when they wish to intimidate a neighbor, or when a mother alpaca wishes to bond with their baby (called a cria).

- **Screaming**- It describes itself. They tend to do this when afraid (can you imagine?). Poorly socialized alpacas will scream all the time at the large naked two-legged apes they perceive is attacking them, so do be sure to bond with them from a young age. Some alpacas are just naturally high strung and will do this whenever they are afraid. It is extremely common for alpacas to scream when they see their young being handled in a way they don't agree to, or when they are being attacked by a predator.

- **Grumbling**- Sounds like gargling mouth wash a bit. You will usually hear this amongst the alpacas communicating with one another. Think of it akin to a dog or cat growl, it basically means get out of my space. They tend to do this when grazing together, and the other alpaca is cutting into its action.

- **Humming**- This is the most common noise you will hear in your time spent amongst alpacas. They tend to hum for many reasons. They will hum when worried, curious, happy, bored, afraid, and stressed out or when something is making them generally apprehensive. You will also hear crias hum nearly constantly when nursing, and when in their mother's presence. The humming will be almost nonstop (well, figuratively speaking) for roughly the first half year of the crias life.

- **Snorting**- Sounds exactly as you would imagine it would sound. This serves the same function as grumbling, just a bit sharper in meaning. It is basically one alpaca telling another to get away.

- **Screeching**- Not to be confused with screaming! Males will screech when in fighting mode, usually over females or territory (what else, eh?). Females will screech (which sounds a little lower in tone) when enraged. If you hear an alpaca screeching it is very, very angry and is a sign of severe stress.

- **Warning call**- Probably the most unique sound alpacas make. It is very shrill and will definitely get your attention. It sounds like an undulating metallic whistle. They only do this when they see something they perceive as a predator. This is meant to warn the rest of the herd, and some of the smarter alpacas have figured out this is a good way to get a guard dog or llamas attention to an animal they want out.

- **Spitting**- More action than sound sounds just like you would imagine. This is always meant to display displeasure, usually towards other alpacas, but occasionally to their human handlers. I guess try not to take it personally?

Alpaca body language

Body language is probably the way alpacas communicate the most. If you learn to read an alpaca's body language, you will be way better off as an owner. Unless you are socially daft, it does not take a ton of experience to quickly get at what your alpaca's mannerisms and stances mean, as they superficially resemble that of most other livestock. There are five general things you should understand about your alpaca's body language.

- **Facial expressions (ear, lips, tail, and head)-** Relaxed alpacas will have its ears up or slightly relaxed. An alpaca that is paying attention to its surroundings (whether stressed, curious, etc.) will have its ears cupped towards whatever it is trying to listen to. You should also consider the tail to be an honorary part of the alpaca's face- it will do you well to understand that the tail is the alpaca's version of a temper thermometer. The higher the vertical positioning of the tail, the more aggressive the alpaca's behavior will be. Alpacas will draw their lips back too, usually when distressed. If you see the alpaca's ears drawn back and its lips pursed, watch out! It's probably going to spit at you.

- **Submissive crouching-** You will see this in young and adolescent animals the most. The alpaca in question will lower its head and curve its neck towards the ground and flip its tail upwards. This is simply a submissive pose it assumes around older and potentially stronger and more aggressive alpaca. This just means the youngster does not want any trouble.

- **Alert Stance-** Usually, alpacas become alert when they see an animal they are not thrilled about seeing. The tail is somewhat elevated, the ears are pointed like satellite dishes towards the critter in question, and it will rudely stare down whatever the perceived threat is. Usually, an entire herd will take this position. Remember, alpacas don't really have the ability to fight, so this pose is usually followed the shrill alarm call described above, or outright fleeing. Admittedly,

51

an alpaca stampede is a cute sight to witness. Do not be alarmed if they strike this pose and you do not see what they are paying attention to- they have excellent eyesight and even better hearing. They are prey animals after all.

- **Aggressive stance-** The animals will ruthlessly stare each other down, ears drawn back, maybe even a little snarl with an upturned head. Usually, this means either a fight is going to break out, or an animal will yield. The next step in escalation is a spitting match- pretty gross. You will most likely see this in your herd with overly defensive mothers posturing to defend their crias from some perceived danger or slight. Yes, alpacas have personalities, so if you know two particular alpacas have it out for each other, they will consistently standoff, and it may be time to separate or move the troublesome alpacas as you see fit.

- **Broadside pose-** This pose is meant for intimidation, pure and simple. Males most often use this pose to signal aggression from a distance away. Head will be turned up, they will stand rigidly and hold their tails up as high as they can muster. Males seem to be standoffish with females when not interested in them sexually and will strike this pose to keep some distance. Conversely, females will strike this pose when they want to signal in as clear and simple terms as possible to potential male suitors they are not in the mood for sex. Also, mothers will strike this pose to keep alpacas they are not happy with far away from their crias.

Alpaca scents

Being herbivorous prey animals, alpacas have an excellent sense of smell. This is used to forewarn the presence of predators, and also to know where food and water sources may be best found in their surrounding environment. Their noses are co-opted to communicate with each other as well.

The most common use of scent is to mark off the boundaries of territories. In the wild, vicuna males will mark their territory off with piles of dung acting like some sort of boundary marker. Males will also sniff the dung of females to determine if she is available sexually. Females tend to use sniffing to determine which cria is theirs. When it comes to alpacas and people, they will smell a human beings face (assuming you get to their level) as a sort of greeting. It really is quite cute.

Additional information

If you should happen to see your alpaca acting in a very unusual fashion (i.e.- not in a way described above) it is important to have a veterinarian see it. It may be suggestive of some underlying malady. Animals have a limited way to communicate stress or discomfort, and it is important to be keenly aware of what it is trying to let you (or its herd mates) know. We will cover common alpaca maladies later in the book.

Chapter 7: Feeding your Alpaca

Eating is probably your alpaca's favorite activity, besides sex. Being grazers naturally, they will constantly eat. Luckily, the diet that is most natural to them is also unlikely to make them obese, but they will readily eat through your budget and then some if you let them. The biggest dangers your alpaca faces from its diet is lack of proper nutrition, or accidentally eating inappropriate or harmful food. Luckily, their dietary needs are not radical and closely mirror that of other common livestock.

Commercially available food

High-quality hay! Far and away the best forage you can give your pet alpacas is long fiber grass hay. If you should give your alpacas high protein hays (alfalfa, clover, etc.), you may run into nutritional problems which we will discuss in a bit. Aside from hay, you should also provide supplemental nutrition, there is no way hay alone will be sufficient fodder to keep your private alpaca herd healthy.

Most supplemental food for alpacas comes in the form of pellets. The pellets generally tend to be a mix of a variety of grains, vitamins, and minerals. Whatever your pasture and hay lack, these pellets will provide for your alpacas. If you are particularly worried about your alpaca's nutritional intake, feel free to mix up a few pellet varieties, and maybe throw in some additional vitamins and minerals. Some owners take it even further and throw in probiotics to aid the alpacas with digestion. If you just go slightly above the minimum, you and your alpacas should be just fine.

Brands that come highly recommended:

- Mazuri Pellets
- Nutrena Feed
- Blue Seal Feed
- Kent Feeds

Ensuring a balanced diet for your Alpaca

Alpacas can be just as varied as individual human beings. They will need a different thing at different stages of development. The smallest baby crias will be nursing from their mothers. Nursing mothers will need supplemental nutrition. Luckily for you, there is almost no guesswork or scientific approximating to be done on your end. All the above-listed brands have special formulations for differing alpacas (especially for gestating and nursing alpaca mothers). Speak with your veterinarian, take their frank advice and buy whatever supplements he or she feels comfortable with providing to your alpaca.

One more point about alpacas- they can be finicky eaters. You will notice when they forage, they tend to go for the softest parts of hay and will often ignore the rest of it. This is nothing personal, it is just to them, and the food is free, why not be picky? You may want to spare yourself some sanity and money by buying a waste-free hay feeder. Alternatively, some pet owners have bought plastic-wrapped bales of hay, and only unwrap the top portion. They continue to unwrap as the alpacas eat their way through the rest of the bale. If you allow them free reign and unfettered access to an entire bale, they will pull it apart, trample most of it, and only eat a tiny fraction. I mean, who wants to eat food that has become muddy once trod underfoot by your neighbor? The exact same thing goes through your alpaca's mind.

Natural treats for your Alpaca

Most grains will be treats enough for your alpaca. Heck, they will be pretty excited about their normal feed. Below is a list of safe treats for your alpaca:

- Carrots
- Apples
- Grapes
- Pears
- Plums
- Cherries

- Blackberries
- Raisins
- Peas
- Raspberries
- Strawberries
- Bananas
- Watermelon
- Squash
- Cantaloupe
- Peaches
- Broccoli
- Green beans
- Cucumbers
- Spinach
- Peanuts
- Crackers
- Sugar cubes

If you have any doubts about what you want to feed your alpaca, be sure to ask your veterinarian. The above list is more of a greatest hits list than a comprehensive end all be all list of what they can eat for a treat.

Vitamins and supplements

A word of advice when it comes to alpacas and salt licks. A lot of alpacas are not actually all that keen on licking a single large block. Some owners will take salt chips and grind them up and mix them in with the regular feed. If you notice your alpaca not paying the right amount of attention to the salt lick, this may help you.

As for supplemental vitamins and minerals, they will need different formulations based on their individual needs (pregnancy, old age, etc.) and the time of year you are feeding them. Luckily for you, the alpaca industry has dedicated mineral and supplement providers to take the guesswork out for you. The most reputable company to get your alpaca specific minerals from is a company called Stillwater Minerals. They seem to have exhaustively researched the dietary needs of both alpacas and llamas and have an exhaustively detailed explanation about the precise ratios of minerals and vitamins in their supplements, and why they are needed. Well worth the read if you do the research on their products.

Dangerous foods

Although not exhaustive, below is a pretty lengthy list of species of plants (and their associated nuts and fruits) you **must not feed your alpaca.**

- Acorns
- African rue
- Agave (as in tequila plant leaves)
- Amaryllis
- Arrowgrass
- Autumn crocus
- Azalea
- Beargrass
- Sand begonia
- Bird of paradise
- Bitterweed
- Black and Mountain laurel
- Black locust
- Black walnut
- Bladderpod
- Black snakeroot
- Bleeding heart
- Blue-green algae (someone somewhere tried this)
- Bracken fern
- Broom snakeweed
- Buckeyes
- Buckwheat
- Buffalo burr
- Burroweed
- Bur sage
- White ragweed
- Buttercups
- Butterfly weed

- Calla lily
- Calamondin orange tree
- Camas lily
- Carnation
- Castor beans (highly lethal to people as well)
- Catclaw
- Chinaberry
- Chokecherry (basically, any sort of cherry, really)
- Christmas cherry
- Christmas rose
- Cocklebur
- Corn lily
- Cress
- Crotalaria
- Crow poison (as if the name were not warning enough)
- Crucifers
- Cyclamen
- Daffodil
- Daisy
- Daphne
- Deadly nightshade
- Deathcamas
- Devils ivy
- Dumb cane
- Drymary
- English ivy
- Elephant ears
- Eucalyptus
- Eyebane
- False hellebore
- Fiddleneck
- Firecracker
- Foxglove
- Geranium
- Ginko tree
- Golden chain tree
- Greasewood

- Groundsel
- Gumweed
- Hemlock (this killed Socrates)
- Henbane
- Holly berry
- Horsebrush
- Horse chestnut
- Hyacinth
- Hydrangea blossom
- Indian hemp
- Inkweed
- Iris
- Ivy bush
- Jack in the pulpit
- Jequirity bean
- Jerusalem cherry (ironically, native to Peru)
- Jimsonweed
- Johnson grass
- Jonquil
- Juniper
- Klamath weed
- Labrador tea
- Lantana
- Larkspurs
- Laurel
- Leopardbane
- Lillies
- Lily of the valley
- Locoweed
- Lupine
- Manchineel
- Mandrake
- Mayapple
- Mescal bean
- Mesquite
- Milkweed
- Mistletoe

- Monkshood
- Morning glory
- Mountain laurel
- Mountain mahogany
- Mushrooms (not a plant!)
- Mustards
- Narcissus
- Needlepoint ivy
- Nightshade
- Oak brush
- Oaktree (acorns and leaves)
- Oleander
- Orange sneezeweed
- Oxalis
- Pasque flower
- Philodendron
- Pin cherry
- Podocarpus
- Poinciana
- Poinsettia
- Poison hemlock
- Poison ivy
- Poison oak
- Poison suckleys
- Poison sumac
- Pokeweed
- Poppy
- Potato plant
- Pothos
- Prince's plum
- Privet
- Pyrocantha
- Rattlebox
- Rayless goldenrod
- Rhododendron
- Rhubarb (which makes excellent pies, just not for your alpaca)
- Ryegrass

- Rubberweed
- Russian thistle
- St. Johnswort
- Sandcorn
- Sesbane
- Silverling
- Skunk cabbage
- Snow on the mountain
- Sorghum (so not EVERY grain is acceptable to your alpaca)
- Spathe flower
- Spurges
- Stagger grass
- Star of Bethlehem
- String of pearls
- Tansy ragwort
- Thornapple
- Tobacco (which is bad for people too)
- Tomato leaves
- Trumpet vine
- Tulip
- Vetch
- Violet seeds
- Water hemlock
- White ragweed
- White snakeroot
- Wild carrots
- Wild cherry
- Wild cucumber
- Wild parsnip
- Wild peas
- Wild plums
- Wisteria
- Yellow Jessamine
- Yew tree

The above list of plants is not exhaustive (despite the good faith attempt)- you should consult your veterinarian for their knowledge as to what plants may be native to your area that may be of concern for your alpacas.

Additional poisonous things

- Walnuts
- Monensin- Common supplement for chickens and cattle, AVOID
- Urea
- Molds
- Blister beetles

Plants that are terrible, though not lethal

- Foxtails
- Stinging nettles
- Poison ivy

The above are extremely irritating plants- to the point you may need to take your alpacas into the veterinarian should they come into contact with them. Stinging nettles are actually quite nutritious for man and animal alike, but they sting like the dickens!

Chapter 8: Play and Exercise

This will probably be the chapter in the book that is the most fun. Although most people never consider the fun you can have playing with your livestock, alpacas are definitely fun to horse around with, pardon the turn of phrasing. It is true, an alpaca does not play like a cat or dog, but they have a sense of fun to them, especially when you are comfortable with them and vice a versa.

Playing with your Alpaca

If you are of a devious mindset, you can always chase your alpacas, but this is somewhat cruel and would sow the seeds of doubt and mistrust of you in your herd. Still, if a neighborhood kid should feel somewhat giddy and give them a good sprint, it is hardly the end of the world. Once you have established a trusting relationship with your alpaca, you can actually train them quite easily; they are said to be smarter than sheep, although not quite as smart as a dog. You can teach your pet alpaca's how to do the following:

- Catch
- Sit
- Litter train
- Train to be a therapy animal
- Tow carts
- Learn colors
- Navigate obstacles

To be frank, the litter training bit is a little overblown. Alpacas naturally pick a spot and use that as their latrine, although if you move their dung pile, you have effectively trained them to follow that designated spot as the latrine area.

Nina Faust, an Alaska resident, demonstrates some of the basic tricks you can teach your alpacas on the Internet. This includes catch and teaching them to ring a bell. Basically, the method of training is that you use operant conditioning. That is basically a fancy psychological term of positive reinforcement- when you see a behavior you like from your alpaca, you reinforce it with a treat. This is a bit time intensive, so be patient. Patience and sugar cubes will go a long way towards modifying your alpaca's behavior.

Perhaps the easiest trick to teach your alpaca is how to sit. Some people call it cushing or haunching, but basically, whenever your alpaca sits on its own accord, use a command word (firmly saying "sit" for example) and give it a treat. Once the alpaca recognizes that following this command will provide it with a treat, it will quickly get with the program. This will also make the alpaca far more receptive to learning tricks in the future, especially if it sees you wandering around its enclosure with a fistful of sugar cubes.

When it comes to towing carts, basically, this means towing a small wagon. Remember, alpacas are not beasts of burden, but a small wheeled cart should be within their ability to haul. Clearly, it may only haul small children, trying to force an alpaca to haul around a fully grown adult human is simply cruel.

Alpaca farmer Terry Crowfoot managed to train her alpacas to distinguish between colors, only using a clicker and some treats. Now, it is not as fancy as it sounds- there were no verbal commands used. Instead, every time the alpaca selected a red plush, the clicker would click, and it would be given a treat. In time, it started hunting exclusively for red plush toys. One can probably extrapolate this over time with verbal cues, so it makes for an interesting experiment you can try with your pet alpaca, should you be inclined to do so.

Finally, the most interesting thing you can probably train your alpaca to do is navigate obstacles (you know, like in the Westminster dog show- though admittedly not with the same kind of vigor and enthusiasm a dog may show). Again, alpaca owner Nina Faust has a very interesting video showing alpacas navigating obstacles. This is probably the most practical thing you can show your alpaca because should you be inclined to hike somewhere with your alpaca, they can easily navigate obstacles (like fallen trees or abandoned car tires) that they would normally be quite skittish of naturally. Simply set up obstacles, give the alpacas a physical cue to follow (in the video it appears to be a piece of cardboard at the end of a stick), and a clicker with some treats and you will be on your way towards training your alpaca to navigate obstacles. As with anything else, time and reinforcement are key.

Common hazards for your Alpaca

Never make your alpaca feel like it is in a situation where it may have to fight or flee. Alpacas are naturally curious animals, use this towards your advantage. Have confident, but not aggressive, body language when training your alpaca. Speak reassuringly and apply treats where appropriate. Just like children, alpacas have their limits when it comes to mental stamina and may simply burn out or get tired of activity- that is ok. They will also be very wary of any new thing in their environment, so do not be disappointed when they view whatever obstacle course or maze you set up with a deep suspicion. You may even want to set up your course a number of days prior to when training will begin simply to acclimate your pets to this new feature. Remember, a panicked alpaca is an injured alpaca.

Toys

Toys are an important enrichment tool for your alpacas. If you lived in a large rectangular field with absolutely nothing to interact with, you too would quickly bore. As with most animals, a bored alpaca is much more likely to become a troublesome alpaca. Enrichment provides the following benefits:
- Increases behavioral diversity
- Increases the frequency of natural behaviors
- Lessens how often abnormal behavior occurs
- Increases confidence and interaction with the wider environment
- Increases the mental resilience of your alpaca to interact with change in its environment

This last point is extremely important- if your alpaca has a "nothing new under the sun" attitude, it will make your pet far, far easier to handle and transport than an alpaca where everything is new and unknown to it. You won't even need a lot of toys to make your alpaca's life more interesting, below are some simple suggestions to improve your alpacas quality of life:

- Offer limbs and branches of safe tree species to nibble on
- Rotate grazing areas periodically
- If there is a mild winter day, allow for different hay feeding spots outside the barn
- Offer a variety of hay, instead of the same hay every time
- Offer oat straw
- Vary the location where you feed your alpacas
- Hide treats around the enclosure
- Create varied terrain (hills, dips, etc.) if it is within your budget and power to do so.
- Provide scratching posts or brushes attached to walls to allow your alpaca herd a simple place to scratch that itch.
- Put mirrors in the enclosure
- If safe- give your alpacas a view. Streams, distant foot or car traffic, etc. are all relatively fun for your alpacas to observe.
- Introduce your alpacas to new foods from time to time.

Popular Toy Options

Two toys you should seriously consider to adding to your alpaca paddock are a treat panel and roll around the plastic jar. The treat panel is simply a long wooden box with little doors your alpaca can manipulate with its lips. It pushes the door to the side to find a treat. The plastic roll around treat jar is even simpler and requires far less carpentry to make. Simply find a clean plastic jar with a screw on lid. Drill a moderate sized hole in the side of the plastic jar and fill the jar with treats. Your alpaca will entertain itself for hours as it rolls the jar around to try and prompt it to release a treat.

Weight loss

Although it is very difficult to achieve, your alpacas can be prone to obesity. It may seem cute to have a fat alpaca, it is actually very bad and dangerous for them. The most common causes of obesity in alpacas are as follows:

- Slow metabolism
- Aggressive eating (especially if asserting dominance over other alpacas)
- Genetic disorders

Long story short, your alpaca's weight can be managed either by changing its diet (easier said than done in both humans and alpacas, especially with a family or herd to consider) or reorganizing the herd (basically, getting the aggressive alpaca out of there). As always, consult with your local veterinarian for the best course of action.

Enclosure features

Life is boring as a livestock animal- it can be much more boring if the owner in question puts in minimal effort. For reasons described above, it is imperative you provide an enriching environment for your alpaca herd. There is a range of things you can do to provide an enriching environment, it all depends on your space limitations and budget constraints. Below are a few suggested features.

Enclosure enriching features

Cheaper features

- Toys
- Mirrors
- Rotating grazing fields periodically (good for the fodder growth as well)
- Providing a view for your alpacas
- Switching brands of food, hiding treats
- Regularly interacting with your alpacas (from casual to time spent training)

Moderate features

- Other farm animals
- Playing music in the barn or grazing field
- Setting up mazes and obstacle courses
- Planting and guarding alpaca safe trees (they will strip the bark if you do not provide adequate protection- bonus, down the road, they will provide another area of shade down the road)
- Planting a garden for the alpacas to ravage at some point in the future. Remember to consult with what plants are alpaca

67

safe, fence it off, let it grow, then one day remove the barrier. They will have a field day, it is a moderate amount of work and money for a one day thrill, but oh what a day it will be.

- Multiple salt licks

Expensive features

- Customizing the barn. This can include predator resistant windows, sound systems, automatic re-waterers, interactive features, heating, and cooling, etc.
- Altering the terrain on your grazing fields. With some ground moving equipment, you can easily create hills and dips that will more closely resemble the alpaca's native turf of Peru as opposed to just an open plain like a field. If you really do some research, find boulders with alpaca safe minerals and drop them in the field as well, this is how they get their minerals in the wild and will provide a nice alternative (NOT substitute) for a traditional salt lick.
- Installing water features- Ponds with a gradual change in depth. Besides providing another source of water, if you properly manage the pond (fish to keep mosquito larvae out, alpaca safe water plants, etc.) it will attract interesting wildlife for your alpaca to observe. Fish to jump out of the water, migratory birds will stop by, a watering hole for the other animals in your enclosure to get in close proximity and interact, who knows? It is basically like installing the town tavern for your alpacas.
- A sand feature. Basically, a sandbox. Make sure you shoo away neighborhood cats, as they may use this as a little box. That is the last thing you would like to wash out of your alpaca's beautiful fleece. Also, make sure the sand is washed before installation- sometimes sand is treated with chemicals that are not ideal for your animals to interact with. Play sand is ideal for this reason, as there is a lower concentration of these undesirable compounds, but still, run a hose over the feature for a few hours. In no time, the alpacas will get the memo and take dust baths from time to time (or simply frolic).

Chapter 9: Hygiene

Hygiene is of paramount importance, regardless if you are an alpaca or human being. Because of your alpaca's generally furry and fuzzy disposition, it will have more grooming needs than, say a pig. The amount of work is not overwhelming, however, and alpacas are generally clean animals. This does not mean they will require zero grooming though, as neglect can be deadly for your alpacas- they are domesticated animals after all and will require significant human interaction from time to time in order to live the healthiest and most fulfilling lives possible.

Alpaca Grooming

Shearing

Grooming with alpacas is synonymous with shearing. Shearing is a pretty hands-on process, luckily though, alpacas will only need to be sheared once a year in most climates. Shearing tends to occur during the summer when the heat is the highest outdoors, for both your alpaca's safety and comfort. You will need to invest in a good solid pair of shearers, for your sake and the alpaca's sake, buy some electric shearers- it will save your wrist a lot of work and the alpaca unnecessary time spent held down and stressed out. If you are squeamish about using shears yourself, there is almost always someone around willing to do it for you for a price; perhaps even your local veterinarian can offer this service.

Down to the nitty-gritty. If you want an easy experience shearing your alpacas, arguably the most important part happens before you even lay hands on them. Maintain your alpaca spaces as best you can- remove as many twigs and noxious vines from the outdoor spaces as possible. Evergreens also provide a never-ending source of pine needles that seem to be magnetically attracted to your alpaca's fleece. Remember, you can put in this work ahead of time, or you can do it all at once when you are shearing your alpaca. It is no fun for both you and your alpaca to have a stop and go shearing process, and the fleece seems to really grasp onto plant matter for its life.

About a week out from the shearing day, you will need to make some adjustments. Try and section off your outdoor paddock to the cleanest spots available with everything the alpaca will need (shelter, water, and a foraging space). You will do well to withhold hay and straw from your alpaca in this time- feed them alpaca pellets for this week. The reasoning behind this course of action is that the intervening week will allow enough time for debris (like straw!) to simply work its way out of the outermost layer of your alpaca's fleece. This will save you a lot of grunt work come alpaca shearing day.

Hopefully, by this point in your alpaca herding career, you have invested time in building a relationship with your alpacas. If not, they will freak. Best to not do this by yourself until you are really, really experienced. Go out into your paddock, and corral an individual alpaca into your preferred indoor shearing space. Most alpacas will have to be held down to the ground, both for their safety and yours. Some people insist on having a huge team of people to hold the alpaca in place, they do sell special shearing tables that you can restrain your alpaca too, some people will take the extra step and lay a mattress on the ground to hold the alpaca against. These are all acceptable ways to shear, and current research does show holding an alpaca down on the ground is the most stressful option, but not by a huge margin. Then shear. There really is no secret to it, it is like an army haircut. Get as much off as you can as quickly as you can.

Now, once the shearing is done, you have two options. You can throw away the fleece, which would be a real shame, or you can hold onto it. It is pretty valuable stuff, and raw alpaca fleece currently sells for $3-$5 an ounce (£2.30-£3.80 per 28 grams). This all varies depending on the local market, laws, and quality of your alpaca's fiber, but do some local research before you bin the stuff. You could be leaving an awful lot of money on the table (or in the trash). Arguably, the fleece is the number one reason most people own alpacas in the first place.

How to use an alpaca fleecing table

So, to be blunt, a fleecing table is basically a table with a rotatable surface. There are many videos on YouTube showing exactly how they work. You escort your alpaca to the table when it is tilted broadside, and you tie your alpaca to it. Once the alpaca is secured to the surface, you rotate it and bam, you have your alpaca held to the shearing table, no fuss. Shearing tables have a wide variety of options (some have a hugging mechanism, etc.) so the price can fluctuate depending on add on's and market. They typically range from $1,000 (£765) to $3,000 (£2,295) in price, so it is a little pricey.

Washing your Alpaca

Try looking for this on the Internet, and you will more likely than not find inquiries about how to take care of sweaters and socks made of alpaca fleece. Luckily, most alpacas actually look forward to the bathing time- more times than not owing to the cooling effect of the water. In fact, most of the time, a light hosing down will more than suffice as an elementary bath for your alpaca herd (and in fact, is the only kind of bath permitted before showing your alpacas in a show). You really should not bathe your alpacas too frequently, but if for whatever reason they get filthy, make sure to buy a livestock friendly shampoo and wash them as you would any other animal.

Hoof maintenance

You will need to periodically trim your alpaca's nails (unless you happen to have them roaming on an extremely large and somewhat rocky paddock, they may trim themselves in this instance through wear and tear). The process for trimming nails is not entirely unlike shearing- namely, you are going to need to restrain your alpaca, and having helpers makes the process far, far easier.

Alpacas do not really have a true "hoof" in the sense that cows and horses have them. Instead, they have two prominent toes (which are actually the 3rd and 4th digit, respectively) with a pad and a nail at the front of them. The pad superficially resembles a dogs foot pads, in both appearance and texture. This soft footing is part of the reason why alpacas are particularly easy on whatever landscape you put them in- their trampling is far more light-footed than that of a cow!

You will need to trim your alpaca's nails roughly every three or four months, although alpacas on softer ground will grow their nails out more quickly. You can restrain your alpaca with helpers holding it to the ground, on a shearing table, whatever works for you. A quick word of advice, alpaca nails are far easier to trim when wet, compared to when they are dry- so you can wait for a passing rainstorm, or simply hose down their legs. Once the alpaca has been secured, get out your hoof trimmer. Trim along the edges and be sure to avoid the quick. If it is your first time, leave a margin of error, and should you strike your alpacas quick, make sure to have blood clotting powder, some gauze, and triple antibiotic ointment nearby, so an infection does not set in. It really is that simple.

Chapter 10: Your Alpaca's health

The most important thing you can do for your alpaca's health is to establish a repertoire with a reputable veterinarian in your area. Usually, your neighbors will point you in the right direction. Be sure your veterinarian is a large animal specialist, as alpacas are considered light livestock. Of course, you will need to take into consideration their enclosure set up and diet as well. Make sure you have a firm understanding of what is and what is not normal alpaca behavior. Once you see something out of the ordinary, try and diagnose it yourself if at all possible, and if that should fail, contact your veterinarian.

Basic anatomy

Your alpaca is a camelid, and thus follows the basic body plan of most camelids. They have a sweet face with large eyes and long eyelashes. They have what can best be described as a rectangular, box-shaped torso. They also have the trademark long slinky neck typically associated with camelids. They also have long legs with a unique foot. You know very well what an alpaca looks like- either you currently own one or intend on buying one. To understand why your alpaca looks, well, like an alpaca, it is worth getting into the history behind the species.

Biological History of your Alpaca
As discussed earlier in the book, alpacas are the domesticated descendants of the vicuña- an animal with a striking resemblance to an alpaca. The only obvious difference between the vicuña and the alpaca is the length of the fur- your alpaca is far fluffier than any vicuña. The ancient forbearers of the Inca civilization in the Andean region of South America captured some vicuñas and then carefully selected them for one trait in particular- maximum fleece production. No sheep are native to South America, so this was the one available source of wool for these native Americans.

A secondary use for the alpaca was for food, and indeed, they are an excellent source of lean meat (having been described as somewhere between venison and lamb in flavor and texture), but primarily, they were used for their fleece. Unlike their cousins the llama, they were never used or intended to be used, as pack animals. This is why they retain the diminutive stature of their ancestor the vicuña. Alpacas were believed to have been domesticated about 5,500 years to 6,000 years ago and are recorded as having moved further down in altitude about 3,800 years ago. With selective breeding, alpacas developed a far wider variety of color schemes than their wild ancestors had.

Anatomical Features most Camelids have

Common Terminology
If you already own horses, you are probably familiar with most of the below terms. If not, here is a review (or translation, depending on how you view it) of some of the most noticeable anatomical features of your alpacas.

- Poll- Tuft of air on the top of the head
- Withers- Part of the back between the shoulders and neck
- Loin- Lower back
- Croup- Hip area on the back
- Tailhead- The root of the tail
- Shank- Lower leg, below the knee
- Hock- Lower part of the back legs behind the knee
- Gaskin- Muscular part of the hind leg between the stifle and hock
- Stifle- Upper knee on the back legs, kind of like an elbow

Of course, there are common sense features too like the ears and eyes, but these do not need any sort of translation for most people. We will go into detail about these other features as well, but I will not waste your time explaining what you should already know.

Interesting Features

- Alpacas have 74 total chromosomes (as opposed to 46 total for human beings)
- All *Camelids* have 74 chromosomes, and thus, can cross breed producing fertile first generation progeny. We will cover these interesting crosses later.

- Alpacas have no upper teeth
- Due to the unique three-chambered stomach alpacas have, their dietary requirements are minimal. They only need to eat 2% of their body weight per day to stay healthy.
- Efficient digestion is aided by storing food in the rumen (the first compartment of their stomach), regurgitating it to re-chew (this is the cud) and re-swallowing it back to the other two compartments in the stomach.
- If you look closely at your alpaca's neck around feeding time, you may see a lump moving its way back up the neck- this is when they bring their cud up. No need to worry.
- Alpacas come in more natural colors than any other livestock species

External Features

This section will be dedicated to explaining the common features of your alpacas. This will help you determine if what you are seeing is normal. It is important to review this, simply because things are not always what you would expect. There are plenty of anecdotes of panicked new owners calling their veterinarians panicking about the rectangular shape of their goat's pupils, so let us hope we can avoid a similar situation with your alpaca.

Eyes

Alpacas have very beautiful, even soulful eyes. You will notice their pupils are rectangular when they are relaxed. There is another interesting aspect of your alpaca's eyes you may not realize. The environment they are from (the high Andes Mountains in South America- in case you already forgot) has very little in the way of natural shade. There is a unique structure called the granula iridica- it basically looks like an extension of the iris into the pupil area. Its sole function is to provide additional shade for the pupil. Alpacas have somewhat of a variety of iris colors, with some alpacas having brown eyes and some having blue eyes, depending on the parentage.

Ears

Alpacas have large ears- they are somewhat cupped in appearance. They can be pointed around towards a sound source they are keen on paying more attention towards. To be frank, the ears are probably one of the best ways to tell your alpaca's mood- so be sure to pay attention to them.

Nose

The alpaca's nose resembles that of a camel, due to their close genetics. The interior of the nostrils are designed to flare to allow for more surface area within the nose. The more surface area exposed, the more air is drawn in, the more receptors there are to interpret whatever odors are in the air. For this reason, alpacas have an excellent sense of smell, and frequently smell animals (particularly those they do not like) before they even see them.

Mouth

The mouth of the alpaca will take a little getting used to. It is perfectly normal for your alpaca to not have any teeth whatsoever on its upper palate- that is just how they come. They look a little like an elderly person when they yawn, some owners find this appearance endearing. Some children have been alarmed by this, so please explain to them that your alpaca is perfectly healthy and normal by its species standards (remember, people, fall into the trap of anthropomorphizing all the time). The teeth your alpacas do have are highly specialized, and each serves a unique purpose. You will start seeing them erupt in your crias as they mature into full-grown alpacas in roughly the second to the third year of life.

Alpaca males have fighting teeth, they are used for exactly what you would imagine they are used for. Fighting teeth would not be a concern if your male was gelded before the age of two. They are small but razor sharp. They were designed to slash at and tear other male's ears and genitals. Simply file these teeth off at the gum line with obstetrical wire. If you have any doubt whatsoever about which teeth are the fighting teeth, please consult with your veterinarian. These teeth are pretty easily distinguished from your alpaca's normal teeth as they frankly look predatory in nature (like a canine) and grow on the side of the jaw exclusively. Do not be lazy about taking these teeth out- alpaca's have been known to castrate each other using these sharp little suckers. An ounce of prevention is worth a ton of cure, you do not want to be stuck with a hefty veterinary bill for surgical procedures. Your male alpaca's will still fight, but at least they won't have a switchblade in their mouths when they do so!

The rest of your alpacas will occasionally need their teeth filed down as well. You may be worried about hurting your alpaca's teeth- I mean, do not be savage about it, but their dental anatomy is quite different to those of human beings. Their teeth continually grow throughout their lifetime, so this is not an optional procedure. Unlike the teeth of human beings, alpacas only have nerve endings in the very bottom of their teeth, whereas we have them throughout. Thus trimming is actually totally painless for your pet alpaca.

You will want to have your alpacas mouth held open, you can use something creative like a dog bone, or a piece of PVC piping, as long as whatever you use does not produce splinters. Even a rope may suffice, just remember that alpacas cannot open their mouths very wide, so never force it lest you risk severe injury to your animals jaw. Then you use a specialized tool called a tooth-o-matic to trim your alpaca's teeth. Some people get away with using a Dremel drill, but if it is your first time, please consider using a more specialized tool or having your veterinarian come out and show you how to properly trim your alpaca's teeth.

Wool
Remember, the alpaca industry sort of collectively decided that alpaca's do not produce wool- they produce fleece. It is a branding distinction if anything, and it is true that saying alpaca's produce anything resembling a sheep's wool does a tremendous disservice to the high-quality reputation of alpaca fleece. At the end of the day, however, it is just animal hair.

Now desired the world over as a luxury item, the fleece was harvested by the native peoples of the Andes for thousands of years. Described as warmer than sheep's wool and lacking lanolin (a waxy material secreted from the oil glands found in sheep's skin) it is perfectly hypoallergenic. It is also water repellent, soft, and durable and is difficult to ignite. You can see why it demands such a high price in the marketplace.

Depending on the breed of your Alpaca, the fleece will have slightly differing properties. Huacaya alpaca's fleece has natural crimp (a term used in the wool industry to describe the degree of bending the fiber has) and thus produces a better fleece for making yarns. Suri alpaca's fleece has no crimping and is more appropriate for the production of woven goods.

Luckily, your alpaca's fleece is probably one of the easiest parts to maintain on your alpaca. Your alpaca does not need to maintain a bathing schedule (in fact, many farmers would strongly advise you against bathing your alpaca's for its fleece's sake, but this is a pet book, bathing is fine), and they only need to be sheared about once a year. Please shear in the summer, do not leave your alpaca without its winter coat during the cold months. Do try and do your best to keep out foreign objects from your alpaca's fleece, as these may work their way deeper into the fleece if ignored and may be painful to remove come shearing time.

Common Color Morphs

Owing to thousands of years of selective breeding, Alpaca's have many, many color morphs. From the rather drab looking vicuña, which comes in a plain beige form, the alpaca has been bred to have over 22 industry recognized color morphs. Far and away, this is the most color variance of any species of domesticated lives stock. Most alpaca's will come in 5 solid colors, and there are roughly 8 easily definable colors for your alpaca's (without delving too much into industry-specific jargon). These are as follows:

- White
- Light Fawn
- Natural Fawn
- Light grey
- Natural grey
- Rose grey
- Dark brown
- Black

Legs

The word of the day is "conformation." This is a term used in competitions to describe the "correctness" of your alpaca's body to species (or breed, depending if it is a Suri or Huacaya alpaca). The legs should appear straight, although some alpaca's front legs do turn out very slightly in the ankle area- this is not a problem. Just as in cows and horses, alpacas may suffer from being bowlegged, cow hocked, sickle hocked, etc. Consult with your veterinarian if your alpaca exhibits any of these features as to the best course of corrective action.

Like many other herbivores, alpacas have two joints per leg below the hip. The joints on the back legs are the stifle and hock respectively. For the front legs, the joints are called the elbow and carpus, respectively. This means that arthritis has the potential to be doubly hard on your alpaca, so always keep a keen eye out for any sort of lameness. The feet of your alpaca is not hoofed in the traditional sense either. They have pads on the bottom of their feet that are similar to the pads you may find on a dog, while they have distinct toenails protruding from the front of the foot. For this reason, alpacas are also more prone to foot injury than more traditional forms of livestock.

Sexing your Alpaca

It is important to accurately sex your alpaca. This may seem frivolous at first- it should be relatively easy to tell which is the boy and which is the girl alpaca- but as we all know from our everyday lives, gender is more than just genitals. Despite looking similar from afar, there are some distinct anatomical differences aside from the genitals to pay attention to in regards to your alpaca. By far the largest difference will be in the behavioral patterns of your alpaca's however.

Male Alpacas
So, there are large differences between an intact male and a wether (castrated alpaca). As with most animals, removing the testes will drop the production of testosterone substantially in your alpaca, and reduce a lot of the stereotypical behavioral problems associated with the sex hormone. Intact males will need a lot of supervision and can sometimes stress out females in the herd with their pursuit of sex, or other aggressive behaviors. Wethers do not suffer from this behavioral problem, and in fact, it is suggested to have a wether in the vicinity of crias when they are nursing (crias being the babies, of course). Crias like the reassuring presence of additional adults around and intact males will probably want more than platonic

company from the mothers of the crias, which can cause some commotion in the barnyard.

On to intact males. They are noisy and horny, to be blunt, but they are not entirely unmanageable. They (referring to all alpacas) do not have a strict hierarchal system when it comes to anything besides sex. Alpacas are fairly egalitarian, there is no fast and hard rule about who eats food first, who drinks first, who gets what amount of time at the salt lick, etc. The resources in their native range were not exactly defensible in the same way a wolf kill would be (which would explain why dogs have the potential to be aggressive during food time)- a salt lick is a salt lick and usually found by the hundreds of tons in the high Andes. For this reason, the behavior was simply never selected for and therefore did not evolve in the vicuna. Intact males do not like to be in each other's presence and will naturally tend to keep a bit of distance between themselves. Remember that alpacas do not have a defined breeding season, so, in theory, they are always competing in one form or another for the affections of the other gender.

So, trying to alter your alpaca's natural behavior will result in the same thing as if you fought against your dog's natural disposition, or a human beings nature- it just will not work. You cannot expect your intact male alpacas to play nice with each other one hundred percent of the time. What you can do is find ways to alleviate points of conflict. For example, when it comes to feeding time, try to feed the males apart from each other. If they are in close proximity, the odds of a fight breaking out increases dramatically. For this reason, you should have ample space if you intend to keep multiple intact male alpacas together. When a fight does break out, you need to let the losing male have a place to retreat to. Unfortunately, there is no consensus on the minimum size of an alpaca enclosure to keep multiple males happy. It all depends on the individual alpaca's temperaments and the topography of the land, although male alpacas will rarely chase a male on the defensive more than 50 meters (54 yards). So basically, plan on having a lot of land for this particular herd set up.

Now comes to awkward conversation about alpaca fights. It may look cute, and indeed it does look like two huge plush toys are having a throw down, but regular (or excessive fighting) is not healthy. Fighting always increases the stress not only of the participants in the fight but the herd. How would you feel if two family members regularly broke out into a fistfight in your home? It would be pretty stressful, wouldn't it? Now, this is not to say ALL fighting is bad, or even preventable, but sometimes two males will just have it out for each other and make each other's lives (and, by proxy, yours) miserable. In this instance, keep note of who the aggressor is- sometimes some alpacas are just born on the wrong side of the bed. When you are convinced that one male is just too much trouble- you have two choices. Get rid of him (which will be difficult, because who wants a troublemaker?) or castrate him. Castration will most definitely slow his roll. Do remember to remove the fighting teeth, as male alpacas will try and attack each other's genitals in particular- that's a mess you don't want to pay for or clean up, trust me.

If both males have their fighting teeth removed, there is no reason to panic when a fight breaks out. Typically, one side will yield, and no serious injuries are likely. It is not suggested you try to intervene, as you may get knocked over during the exchange of blows, or, even grosser, take an alpaca loogie straight in the face. There will be a lot of spitting, noise, and huffing and puffing, but serious injury is rare. You will see that after a fight, the male alpacas may come down with a case of what is colloquially called "sour mouth"- the phlegm, cud saliva mixture they make and spit at each other is so gross even they cannot really stand the taste, so their lower lip will hang loose and they will drool a bit. All the more reason to stay on your male alpaca's good side!

Nonsexual Features

- **Fighting teeth**- Between the premolar and the lower incisor, you will see what appears to be canine teeth. These exist for one reason- to make other males lives hell. Get rid of them, refer to the earlier chapter on alpaca mouths.

Sexual Features

- **Testicles-** Lift the alpaca's tail and look below the anus, and you should see a familiar sight. You will see two testicles, surprisingly human-like in appearance covered in fine hair. Unlike in human beings, the testicles are symmetrical in appearance, being at about the same depth of droop.
- **Penis-** This is where you enter unfamiliar territory. Aside from the most superficial phallic nature of the penis, an alpaca's penis does not resemble a human penis. It looks a bit like an extended snail with a white corkscrew at the end. This corkscrew feature is called the cartilaginous process and is used to dilate the cervix (presumably by tickling it in just such a fashion) and enter directly into the uterus during sex. The alpaca male is, if anything, very direct. Also, due to the anatomy of the alpaca penis, they urinate backward between their legs, since the tip of the sheath points in that direction, and, like humans, alpacas rarely urinate erect.

Female Alpacas

The ongoing joke in the alpaca world is if you think alpaca's are quiet, then you own all-female alpacas. As with most species of livestock and pets in general, the females generally tend to be easier to keep.

Nonsexual Features

- **Teats-** While in (some) human cultures, breasts are considered inherently sexual, they are not devoted to the purpose of reproduction. Rather, they are to feed the results of sexual reproduction, in your alpaca's case, the crias. Alpaca's have four teats from which to feed on.

Sexual Features

- **Vagina-** Under the anus, works like the vagina of any other placental mammal. Latin for sheath, it is lubricated with a mixture of blood plasma and mucus to ease the penis's entry. Note that alpaca ejaculation proper does not occur in here, like in human sex, rather, the alpaca's penis protrudes directly into the uterus, all the way through the cervix. The orifice of the alpaca's vulva (the externally visible part) is about 3cm long (1.1 inches).

- **Cervix-** Although not readily visible, you still need to know about this feature of the female alpaca's reproductive tract. The cervix is rigid and (relatively speaking, of course) dry when the animal is pregnant. During advanced pregnancy, there will be a mucus plus protruding from the cervix into the vagina. This is a sure-fire sign of pregnancy in your alpaca.

Vaccinations, spaying, neutering and breeding your Alpacas

How do vaccines work

Vaccines are extremely important for your alpaca. This falls in line with the general philosophy of an ounce of prevention is worth a ton of cure. Remember, that vaccines are a *statistical solution* for disease problems with animals- they are not guaranteed to be 100% effective. All a vaccine does (in some cases, precipitously) is reduce the odds of an animal contracting the bacteria or virus it has been vaccinated against. Every individual animal has a unique immune system, and sometimes despite efforts, a vaccine will fail. The true genius of vaccines is when you vaccinate an entire population.

If an animal does not come in to contact with a particular bacteria or virus, it will not become infected by it. Simple enough idea, right? Let us expand on this. Diseases need a certain amount of susceptible animals to spread. If an infection fails to catch in an animal, the animal will not be able to spread the bacteria or virus in question (in fancy terms, the animal will cease to be a vector for disease). So, there has to be a minimum viable population of infect-able animals (or people, in the case of human vaccination) in order for a disease to spread. If 80% of animals can become infected and only 20% have a natural immunity, chances are pretty good a disease will spread to all 80% of the animals simply because it will have enough hosts to spread from. You have to remember, animals do not practice hygiene (other than what we impose on them) and very often do not have the option of not spending time in each other's company.

The longer the duration of a diseases presence, the more victims it will catch. It is the difference between your neighbor having the flu and two of your children coming home with it. The neighbor you will only have passing contact with, so even if you are totally susceptible to the flu for whatever reason, chance plays a big part of whether or not you contract it. Now if your kids come home already infected, you will be spending WAY more time in their presence, to the point of you contracting it becomes a statistical certainty.

So, hypothetically speaking, let us say that a vaccine drops the chance of an animal contracting a particular disease from 80% to 20% (in real life, vaccines tend to drop infection rates from a percentage to 1 in 100's or 1000's odds of contracting). It is not a very effective vaccine we are talking about here. Let us say you have 50 alpacas. Previously 40 of them could catch a disease, but now only 10 of them could. 5 come down with a disease, but then the disease stops dead in its track. What happened? Well, the other 5 who were susceptible could only (in a perfect world scenario) spend a maximum of 10% of their time in the presence of sick alpacas (only 5 had been struck with it, the other 45 themselves included did not harbor the offending pathogen). Had the herd been unvaccinated, there would have been 40 susceptible alpacas instead of just 10, and those first five to contract the pathogen would be spending 80% of their time at least (40 susceptible out of a population of 50 alpacas total) in the presence of either an infected or a susceptible alpaca. The sick alpacas would spend a whopping 700% more time with susceptible alpacas than they would have otherwise.

So the long story short about this is, even is a vaccine is not a silver bullet, herd immunity is important for preventing the spread of disease in your alpaca herd. If a vaccine has totally failed an individual animal but has worked with all the animals around it, that susceptible animal simply does not contract the disease, simply because the disease cannot reach it. Please, please, please vaccinate not just your alpacas, but all the animals (humans included) you can- the math and research bears out that it is one of the smartest ways you can easily increase lifespan for both yourself and your pets.

Vaccinations and Camelids

Aside from vaccination, practicing biosecurity, herd health checks, keeping enclosures clean and seeing to the general well-being of the animals that come in to contact with your alpacas will go a long way towards ensuring their good health. That being said, there are no vaccines currently on the market that have been developed specifically for alpacas (or *Camelids* in general)- this is simply owing to the small market size for biopharmaceutical manufacturers of these species. All vaccines prescribed for *Camelids* (including your beloved alpacas) are therefore used in an off-label sense and were originally developed for other types of livestock like cattle, horses, goats, etc. Basically, even if you have the option to vaccinate your alpacas yourself, you should consult with your veterinarian to see what the best practices and procedures are for vaccinating your alpacas. There are no off the shelf solutions for dosing and prescribing the right brands of vaccines for your alpaca, so please consult with a professional first.

Different classes of vaccines

- **Modified live vaccines**- Contain portions of the virus or bacteria. It is designed to cause a minor infection as opposed to a full-blown infection. This type of vaccine should not be used in alpaca's because they were initially designed with a particular species in mind, and that species was not an alpaca. Using modified live vaccines in cross-species can cause a stronger than anticipated immune response, and therefore, be dangerous.

- **Recombinant vaccines**- They genetically modify a bacteria (most of the time, sometimes it can be a protozoa or something, but that is minutiae) to produce proteins that are identical to that of the pathogenic species. It is basically like training a bloodhound. Rather than bringing the entire pelt of an animal (like modified live vaccines), you bring some hair (recombinant vaccines proteins) to the bloodhound (the immune system) to sniff out the fugitive you want to catch and bring to justice (the pathogen in question). It is a relatively new class of vaccines, so consult with your veterinarian about their efficacy in *Camelids*

85

- **Killed/Toxoid vaccines-** The pathogen is killed (usually with a chemical agent, or radiation), put in a suspending agent and then administered as a vaccine. Basically, it presents the corpses of the diseases to the immune system of the alpaca, and this allows the immune system to easily recognize the pathogen in the future. This is the safest category of vaccine (even safe enough for pregnant alpacas), but this means it is also the vaccine type that produces the subtlest immune response. For this reason, after the initial vaccine, you usually have to provide booster shots a few weeks later to your alpacas (which is all sorts of fun, as you can imagine), and then after *that* about once a year to keep immunity up to date. Usually, the alpacas will develop an administration site reaction in the form of bumps, usually caused by a reaction to the preservatives found in the vaccine.

Some general tips about vaccinating your alpacas
Alpacas that are experiencing a time of stress either because of a big move, travel, a raw deal with another herd mate, recent illness, etc. should not be vaccinated. Stress compromises an immune system, so administering a vaccine, which in of itself is not dangerous, may not have the desired effect. The immune system is maybe so busy responding to stress or other ailments caused by the stress that it may not respond to the vaccine. This will provide you with a false sense of protection, and may ultimately lead your alpaca to the path of illness down the road.

Final note- colostrum may interfere with vaccinations. What on Earth is colostrum you ask? It is mother's milk! With placental mammals, a mother's immunity is transferred through a special kind of milk called colostrum- it has many antibodies in it to strengthen the crias immune system. Unfortunately, these antibodies will interfere with the vaccine itself- so while it is a good idea to vaccinate your crias, you will have to follow up with additional booster shots to make sure the crias immune system got the memo.

Vaccinations

3-way or CD/T Vaccine

- *Clostridium perfringens* type C, D, and *Clostridium tetani*
- *C. perfringens* is associated with diarrhea and sudden death in crias and adults alike
- *C. tetani* causes tetanus (which can be caught as in humans, through wounds)

Clostridium perfringens Type A Toxoid

Unknown if safe for pregnant alpacas, was originally a killed vaccine developed for use in cattle. This vaccine is optional and should only be administered when the pathogen in question is known to be present at your farm or residence.

Other Clostridial Vaccines

It may be used in substitution for the CD/T vaccine. The preservatives in this vaccine have been associated with lumps at the site of administration for alpacas.

West Nile Virus

Alpacas are considered low-risk animals for West Nile Virus, and usually, develop their own immunity to the virus should they be exposed. Still, if you wish to spare your alpaca the pain that is adjusting to the West Nile Virus, you should administer this vaccine prior to mosquito season, with additional boosters as needed.

Leptospirosis

Since this vaccine was developed for cattle, and there are many different species of *Leptospira*, vaccination may not provide 100% protection. Still, something is better than nothing. You will need to provide boosters about 3 or 4 times a year, and prevention usually means limiting contact between your alpacas and neighborhood dogs and rodents, as well as limiting contact with standing areas of water.

Rabies

Yes, alpacas can catch rabies, and it is a very painful and slow death if they do so. Please have your veterinarian apply this vaccine- large animal species usually need a different type of rabies vaccine than small animals do. Also, proof of vaccination for alpacas, unfortunately, may not be enough to spare your alpaca quarantine or euthanasia should it be proven it indeed was exposed. This is due to stringent laws regarding how livestock is to be processed.

Bovine Viral Diarrhea Virus
- Originally developed for use in cattle, this disease is under control in most of the developed world, so ask your veterinarian if it is truly needed.

Coronavirus
More annoying than deadly, it leads to bouts of extreme diarrhea. It is similar to the norovirus that breaks out on cruise ships on infects hundreds of humans at a time, leading to the cancellation of cruise ship tours- not fun to have, extremely contagious, and a long viral shedding period means you will have your work cut out in terms of clean up.

Spaying and neutering

So you have decided that you want to fix your alpacas- congratulations! It is a responsible decision, as in many places, there are more alpacas than homes for alpacas. Maybe you have a troublesome male who needs a quick behavior adjustment- this will work for that as well.

Spaying

To be honest, this is rarely done, but it is a straightforward procedure. Females do not have the same behavioral issues that males do, and there frankly are not a ton of stray alpacas you have to worry about knocking up your females. There is no way for you to do this as an amateur, so do not even think about it. All you can do is keep your alpaca from eating about 12 hours in the run-up to the surgery. Long story short, you (and by you, I mean, the veterinarian) open up the alpaca's abdomen in the vicinity of the udder (usually, just to the front), find the uterus and ovaries and cut them out. Sometimes, it is not necessary to remove the uterus (ask your veterinarian if this is necessary) and just the ovaries (which produce the egg cells that lead to pregnancy) will be removed in a procedure called an ovariectomy. Yes, your alpaca will have to undergo anesthesia for this procedure, so expect a little loopiness afterward. Be sure to attend to the surgical site afterward following your veterinarian's instructions and keep a sharp eye out for any signs of infection. Abdominal infections are a life-threatening emergency and need to be caught early and quickly, and dealt with decisively. Be sure to ask your veterinarian if there is the option to perform the spaying operation with the use of laparoscopy to minimize wounding on your alpaca, instead of the traditional gore fest of open surgery.

Neutering

Castration. It is a heavy word, but good thing livestock do not speak. You may actually want to wait until your alpacas are a little older before performing castration- after puberty (18 to 24 months of age). Breeders like to castrate at a young age so the wethers can be sold as pets sooner, but castration prior to puberty is associated with bone plates not fusing correctly, which down the road can lead the poor male to have severe issues with arthritis due to its altered posture. The castration procedure itself is very similar to that of any other animal- the family jewels will be removed in their entirety. There are two types of castration- scrotal castration where they pop the skin sac open and get the testicles out of there that way (usually done in horses and pigs), and prescrotal castration, done at the ventral base of the scrotum (sparing the sacs direct trauma), a small incision is made and the testicles are pulled out from here.

Once the testicles have been removed, you need to do some postoperative care. Confine the castrated alpaca to a small pen for a day or two (unless the procedure was a prescrotal castration). Complications are uncommon, but keep an eye out for any signs of infection (this includes fly larvae, yuck). Should infection of any type set in, it can be devastating to the alpaca.

Common injuries

Alpacas are fundamentally farm animals. They will do weird things and get themselves caught up in weird situations and get injured. More often than not, they will injure each other with their antics.

Delicate bone structure, long necks and legs and a prominent jaw all add up to a uniquely fragile animal. Shearing also presents a real opportunity for skin injury as well, if your technique is not appropriate.

Common Minor Injuries

- **Shearing injury**- You need to be gentle when shearing. If the angle of the shear is too steep, you can easily cause the teeth of the machine to puncture the skin. Alpaca's have soft and delicate skin.

- **Skin punctures**- This is especially common when alpacas have a lot of stick materials in their fleece. Antibiotic ointment and periodic cleaning of the fleece should more than safeguard against this.

- **Eye trauma**- Alpacas are curious, and they explore the world by shoving their faces wherever. This does not always end well for the alpaca, and they may poke their eyes, especially when browsing in trees or over fences. If you notice an eye injury, please call your veterinarian before infection sets in and the poor animal loses an eye.

- **Torn or broken nail**- This is especially common during nail trimming. Apply anticoagulant powder to stem the bleeding, and you may want to wrap the foot in some gauze for a day or two just to keep the infection from setting if (though, admittedly, this is overkill if you just apply some antibiotic ointments). Of course, if your alpaca somehow manages to lose an entire nail, please call your veterinarian as this will result in severe bleeding, and the risk of infection goes through the roof.

- **Ear nicks**- Occasionally caused during fighting amongst each other, or being attacked by birds. Sometimes ear nicks can be caused by brushing up against hostile foliage or fencing as well. Nothing a little antibiotic ointment can't help.

Common Serious Injuries

- **Torn scrotums-** Your male alpacas will tear their scrotums if you do not remove their fighting teeth. 100% guarantee. Alpacas only fight using low blows. They may bleed to death if this happens in your absence or overnight.

- **Cruciate ligament injuries-** Just like your favorite human athletes, ligaments can sometimes tear in your alpaca's legs. There is nothing you can do about it, you have to take your alpaca to the veterinarian for surgical intervention.

- **Neck injuries-** Happen during fights, or, if the alpaca happens to get their head stuck in something and panics. If not treated promptly, such injuries can easily become chronic. Also, over the long term, should the injury be serious enough (such a broken vertebrate), there may be fusing that occurs. This will severely impact your alpaca's quality of life.

- **Broken legs-** As with any livestock, this is an extreme emergency. You must take your alpaca to the veterinarian (or even better, have your veterinarian come see your alpaca) as soon as possible. Besides being excruciating for your pet, a severe enough break may release bone marrow into the alpaca's bloodstream and cause a fatal clot. Nerve damage may also result from splintered bones. Luckily, unlike a horse, a broken leg is not an automatic death sentence. Since alpacas are relatively lightweight livestock, they can recover- albeit slowly. The vet will come, anesthetize the animal and put on a weight bearing cast. Your alpaca will walk and look funny for a while but will recover in most cases. Broken legs are usually caused by rough fighting/horseplay or a mineral deficiency causing weak bones.

Broken Bones in your Alpaca
Aside from leg bones, alpaca's can pretty easily break other bones. They are most likely the smallest barnyard animals you can own, and so, have a fragile build. Simple horseplay can break ribs- fighting much more regularly so. In the case of broken ribs, all you can do is

damage mitigation. Make sure the break is not so severe it can risk puncturing an internal organ (or, of course, have your veterinarian do it) and give your alpaca some gentle painkillers to ease the healing process. Any broken bones in the neck or leg area should be treated as an emergency, however.

Animal Bites
Animal bites are why you need to vaccinate your alpaca against rabies. Rabies is absolutely fatal to alpacas, one hundred percent of the time without exception. Beyond simple vaccination, you should try your best to prevent hostile animals from having access to your alpaca's. This includes misbehaving male alpaca's- if you have an overly aggressive male alpaca, their fighting teeth can break skin just as easily as those of a large cat. Of course, you can prevent intraspecies bites just by removing these fighting teeth without exception. As for other types of animal bites, there is some general advice.

First, assess the severity of the bite. How deep is it and where did it bite? Bites to the neck can be fatal, as so many important structures are found there (carotid artery, windpipe, spinal cord, etc.). Bites to the leg can be equally dangerous, due to major arteries being present as well as many delicate joint structures. Once you have determined how severe the bite is, determine if you need veterinary intervention. Anything surgical will need a veterinarian. If the bite requires stitching or less though, you can probably get away with treating it on the homestead (presuming you are not squeamish). You may want to get a helper, as your alpaca recently went through a predatory attack and may be extremely skittish. The last thing it wants to relive is something sharp puncturing its skin. Clean the wound out with either hydrogen peroxide or isopropyl alcohol (which will definitely cause your alpaca to buck), and apply antibiotic ointment. Sew or bandage as appropriate. You should probably call your veterinarian over as soon as possible to determine the right course of antibiotic treatment to keep your alpaca from being infected. Animal mouths can be filthy, with predator mouths being doubly so.

When to see a Veterinarian

As often as possible. Your veterinarian is a source of wisdom when it comes to animal care, and chances are very, very good that whatever problem you are running into currently with your alpaca they have seen before. Also, being on a first name basis will help you get quick service- when you have an animal that can potentially bleed to death on your watch, it is very nice knowing reliable help is only a cell phone call away. Whatever you may read, either here or elsewhere (and that includes the entirety of the Internet) should be subservient to the advice of a reputable veterinarian. If ever in doubt about your veterinarian's advice, just as with a human physician, seek out a second opinion.

What veterinarian schedule should my Alpaca be on?
This depends on the stage of life your alpaca is in. Ideally, the moment your alpaca is born, it should be greeted with a veterinarians presence. Once your cria has been welcomed into this world and assessed to be in good shape, it will need to be placed on a vaccination schedule for roughly the first two years of its life. Young animals are always particularly susceptible to illness, and thus need the most supervision of any stage in life. Major first milestones you may want a veterinarian's supervision for will include the following:

- First shear
- First tooth trimming
- Pregnancy
- Spaying
- Neutering
- First hoof trimming
- Euthanasia

The firsts listed above are more for your education, clearly, after a long enough period of alpaca ownership, you will get the hang of regular maintenance.

When is it appropriate to worry about your Alpaca?
Whenever you notice a major behavioral change or change in physical demeanor.

- Unexplained weight loss
- Suspect pregnancy
- Troubled labour
- Repetitive behaviours
- Constant whining
- Lameness
- Limping
- Trouble eating
- Trouble standing
- Sustained aggression
- Abnormal neck function

The above list is clearly not exhaustive, but generally, follow your gut. Once you have owned alpacas long enough and get a hang of what is considered normal behavior, you will notice when things seem off.

Symptoms of your Alpaca being sick

The biggest red flag is a loss of appetite. Alpacas love to eat, one can even say they live to eat. Most alpaca related illnesses involve their gastrointestinal system somewhere along the line, so keep an eye out for that. Diarrhea is not normal- alpacas defecate in pellets, so diarrhea can be indicative of a parasitic infection. Weight loss is bad as well. You may also notice your alpaca's stamina declining- where once you had a perky interested alert animal, now the animal seems morose in mannerisms and capability. This should really set off alarm bells, and you should call your veterinarian to get to the bottom of the problem. It could be something as simple as a nutritional deficiency, or as severe as cancer- you just won't know without a proper medical intervention.

First Aid for your Alpaca

Alpacas, like any other pet, can get themselves into mischief. Sometimes this mischief can be pretty severe, so it is a good idea to have the resources you will need eventually on hand to react to a crisis. Time spent looking for medical supplies is the precious time you really should be spending mending to your injured alpaca. In a medical emergency situation, time saved could very easily be a life saver as well. Be proactive with your animal's health, you know very well as a pet owner it really is a matter of when not if.

First Aid Kit

For simplicities sake, try to keep the first aid kit in one place rather than spread out in a nightstand or something like that. A physical, one-stop shop box is ideal for responding to a crisis situation. These items should be readily on hand.

- **Rubber gloves**- This is more for you than the alpaca. There is no literature to suggest alpacas having any sort of allergy to latex, but if you wish to err on the side of caution, buy latex-free gloves.

- **Catheter syringe**- Makes administering nasty tasty medicines way easier for you as the owner.

- **Collapsible water bucket**-This will allow you to get water to your alpaca quickly. This is important should you suspect one of your alpacas is suffering from heat stroke.

- **Fly mask**- Stress is a killer. If your alpaca is incapacitated for whatever reason, the last thing it needs to be doing is trying to shoo horseflies from its eyes.

- **Spit mask** (for your unruly alpacas)- An injured alpaca is a stressed alpaca. Stressed alpacas quickly turn angry. It is not anything personal, but it is a defense mechanism. For your sake and the sake of your veterinarian, have a spit mask at the ready to save all personnel the effort of having to clean themselves after coming in to contact with alpaca phlegm.

- **Travel trailer**- Sometimes, your veterinarian cannot reach you. You will have to go to him or her to get the services you need. If your regular veterinarian is out of town, and you need to drive to the next nearest one, having a trailer on hand will make the whole process way smoother. Plus the last thing you want is a panicked alpaca inside your car, or a bleeding/diarrhea/phlegmy one.

- **Electrolytes**- Why on Earth would this be in an emergency supply? Well, frankly, if your alpaca is overheating, it may be suffering from a temporary electrolyte imbalance. Electrolytes also work great as impromptu mood stabilizers-kind of like giving a child a lollipop at the pediatrician's office. You have to have something to bribe your alpaca with, and a tasty salty treat is perfect for grabbing their attention.

- **Equine wound spray**- This should speak for itself. If your alpaca should ever sustain a severe open wound, rather than trying to scramble for an antiseptic solution, just have a spray bottle to run to and grab. It will spare you an awful lot of work under pressure, and early treatment is key in infection prevention.

- **Triple antibiotic ointment**- Apply after spraying antiseptic. It will be another layer of defense against infection. Bacteria love nothing more than an open wound to colonize.

- **Blood stop powder**- Should be pretty self-explanatory. Obviously, this powder has its limits, but for minor cuts and such, it is great to have a way to stop the bleeding quickly. Especially if you have to move your alpaca for any reason, it is nice not leaving a tremendous blood trail everywhere.

- **Veterinarian wrap**- After treating a wound, covering it is essential.

- **Terramycin ophthalmic ointment**- The first line of defense should there be an eye injury.

- **Stethoscope**- Need to be able to listen to your alpaca's heart to make sure the ticker is functioning as needed.

- **Digital thermometer**- How else are you going to know conclusively if your alpaca is overheating or has a fever?

- **Livestock scale**- Does your alpaca look thing or is it just your imagination? Know for sure!

Medicine and Food

Food

- Forage (grass in the field)
- Hay
- Alpaca feed
- Occasional treats (sugar cubes, mints, fruits, vegetables)
- Alfalfa (sparingly- it has too high of a protein content for regular consumption)
- Grains (sparingly as well- high protein and calorific)

Medicine

You should ask your local veterinarian what you should keep on hand in terms of readily accessible medicines, as your specific alpaca herd's individual needs will be determined by geography as much as anything else. Alpacas in remote areas are more likely to suffer animal bites, alpacas in tundra environments may get frostbite, alpacas in tropical zones may suffer from heat stroke, etc.

- **Naxcel-** Broad spectrum antibiotic, available only by prescription (all the more reason to be chummy with your local veterinarian).

- **Oxytetracycline-** Broad spectrum slow release long-lasting antibiotic, widely available over the counter.

- **Penicillin-** Extremely common antibiotic, sort of a catch-all at this point in medical history.

- **Tetanus Vaccine-** Discussed at length during the vaccination portion of this book- Alpacas susceptible to *Clostridium* *perfringens* type A and a vaccine for that is not usually readily available.

- **Equine West Nile Virus Vaccine-** It's expensive and needs boosters, and really should only be used after referring to your veterinarian. If your area does have an endemic West Nile Virus, it might be good to have the boosters on hand just so you do not have to pay for multiple veterinarian visits.

- **Lepto Vaccine-** Ask your veterinarian if you have a leptospirosis problem in your area first.

- **Sulfadimethoxine-** Medicine with multiple uses. Although originally developed with cattle in mind, it is used to treat many illnesses. It is used for treating urinary tract, respiratory and soft tissue infections.

- **Amprolium-** Most commonly found in the poultry industry, amprolium is an antiprotozoal agent meant to disrupt the life cycle of (most often associated with) water borne parasites. You can put it directly into your alpaca's water supply, or you drench your fields with a dilution containing this compound to get the little suckers while you are ahead.
- **Ivermectin-** Used to remove parasites- usually associated in humans with treating roundworm infection, scabies and pubic lice (ouch!), more often used in Alpacas to get rid of intestinal worms.

- **Doramectin-** It literally does the same thing as ivermectin, so stocking both would be redundant. Comparison shop to make sure you are getting the best pricing.

- **Fenbendazole-** Another deworming agent, usually sold in paste form. No need for a prescription it is sold over the counter both on the Internet and at agricultural feed stores. Effective against roundworms, hookworms, whipworms as well as the waterborne parasite giardia. Use caution when treating alpacas you suspect are pregnant, as there has been a correlation between using this compound and fetal death in cattle and sheep. As always, discuss with your veterinarian first.

- **Levamisole**- Sold as a soluble drench powder, it is used to treat parasitic worm infections, most commonly used against ascariasis and hookworms. Use caution on suspected (or confirmed) pregnant alpacas.

- **Moxidectin**- Sold both as an injection and as a drench, this compound is used to combat heartworms and intestinal worms.

- **Albendazole**- Another antiparasitic compound, it is usually sold in drench form for agricultural purposes. No need for a prescription, available online. Combats giardia, pinworm disease, and many other protozoan waterborne illnesses. Notably more expensive in the United States than elsewhere.

- **Praziquantel**- Usually sold in combination with ivermectin, it comes in paste form for agricultural purposes. Useful to combat schistosomiasis, clonorchiasis, opisthorchiasis, tapeworm infections, cysticercosis, hydatid disease, and other fluke infections.

- **Activated charcoal**- This is extremely important to have readily on hand for emergency first aid purposes. Activated charcoal is used in suspected cases of poisoning (which means, after feeding it to your alpaca, you have the vet come over as soon as possible to further stabilize the situation) and to treat upset stomach (not so critical, but nice to have). This is one thing that can make the difference between life and death in your alpaca. Please familiarize yourself both with the list in this book of potentially lethal plants for your alpaca, and consult with your veterinarian about what plant species are endemic in your area that may cause a problem for your alpaca.

- **Probiotics**- Helps your alpaca's stomach. Beneficial microorganisms that aid in digestion, nice to have if you have noticed your alpaca having a tough time after dinner.

- **Electrolytes**- It usually comes in paste form, this is great to have during hot weather when your alpaca herd may be in heat distress, or in danger of dehydration.

- **Epinephrine**- For emergency use for the prevention of anaphylactic shock, available only through prescription. This has become notably more expensive in the United States as of late.

- **Thiamin**- A vitamin B compound, this will help your alpaca during times of digestive trouble.

For Evacuation and Emergency Travel Purposes

- **Travel trailer**- If your alpaca is seriously injured enough that it needs to see a veterinarian, it is important to have the means to transport it.

- **Water**- A few spare 5-gallon cans of water are always handy to have, in case you need to drive a long distance in short order (like a wildfire, or another natural disaster). Your alpacas will need to drink during times of stress to stay healthy.

- **Alpaca feed**- You will never regret having an extra bag at hand.

For Sanitation Purposes

- **Gloves**- To keep you from spreading an infection if you are handling multiple alpacas during a day.
- **Iodine**- Great for cleaning out wounds and such.
- **Rake**- Use this to attack the neat pile of poop the alpacas like to grow.

Chapter 11: Young and Old Alpacas

Puberty is a funny time in everyone's lives, and alpacas are no different. One day your beloved baby cria will grow up, and start taking an interest in the other alpaca's in the field- no need to worry, this is all part of the normal growing process for your pet alpaca. Luckily, there is no awkward explaining about how they got here, or what may be expected of them from potential future maters. No, no, no, alpaca's are animals and know exactly what to do one the urges set in. The only major complication with alpaca sexuality is the aggressive nature of multiple males housed in close proximity- that seems to be a problem across a lot of species. Take the fighting teeth, they should be reduced to looking like fighting plush toys in your field. Fail to take this step, and you may have an impromptu castration to deal with.

Things to consider
Please remember, that unlike the vicuña, the alpaca has no distinct breeding season. Sounds great right? Well- not so fast. Alpacas are what is termed in the world of biology an "induced ovulate." This means they will not ovulate unless mated with. So, for the first time, you may want your veterinarian supervising the situation. A sure-fire way to tell if your alpaca is ready for breeding is with an ultrasound directed at the corpus luteum. The mature follicle can be seen, and depending on its size and positioning, can indicate when the best time to mate an alpaca will be. No worries, the presence of alpaca semen will cause your alpaca to ovulate.

Male Sexual Behavior
Male alpacas are pretty stereotypical. Once they hit about two and a half to three years of age, they will become interested in sex. They will be unable to breed until their testicles have grown an appropriate amount and the penis is free of attachments within its sheath. At this point, the penis becomes far more visible and hangs freely- that is when you know he is open for business. Beyond that, you will need minimal effort to encourage your male alpaca to engage in sex- in fact, you will spend far more time controlling with whom and when to have sex than anything else.

Female Sexual Behavior

Female sexual behavior is, of course, somewhat more subdued. Female alpacas reach sexual maturity at a somewhat younger age than male alpaca's do- at about 10 months to 18 months of age. As usual, the females mature more quickly behaviorally as well. Female alpacas are not nearly as noisy and raucous as male alpacas and generally behave better. The only suggestion that a female alpaca is willing to have sex is when she cushes- the fancy alpaca term for crouching and sitting. Basically, female alpacas will do this all the time, so males will constantly be harassing them in order to convince them to have sex. No worries, if the female is not interested and legitimately only wanted to sit, she will either buck the male off or, more to the point, spit in his face. If only dating could be so simple for all species.

Fate of the Young

So sex leads to babies (well, crias in this case)- great! But what to do with that information in regards to your alpacas? Well, unfortunately, alpaca's have pretty complex genetics- so it is very difficult to forecast with any certainty what the crias will look like or how their fleece will be (indeed, *Camelids* have more chromosomes than we do). However, there is a service that will help you determine the appearance and where your entire herd will head in terms of appearance and fleece function. The service is provided by the Alpaca Owners Association and is called the Expected Progeny Differences. You submit some basic information, as well as some physical materials from the alpaca's in question and they can help you manage your herd for its desired result, for a cost, of course.

Anyways, back to the cria, rather than its traits. Gestation is 11 months and, usually, produces only 1 cria. Alpacas have the unique ability to birth within a certain time window- between 9 AM and 2 PM. It is surmised that this unique feature of alpaca birthing evolved when the ancestral species, the vicuña roamed the Andes Mountains. In fact, they still do, however, the vicuña birthed during the brightest hours of the day in order to give their crias time to dry off before the cold mountain night settled in. Cold can be lethal to young crias, so this drying time was important to allow time for the fleece to fluff up and protect the cria in question. Pretty neat.

You should have your veterinarian on hand for the arrival of any new cria. Crias are usually born sloppy looking (like most babies) and weigh between 14 and 20 pounds). They will stand within the hour and nurse not long after getting on their legs. Because alpacas are gregarious herd animals, they will all want to look at the new arrival. This is perfectly normal, in fact, desirable for proper socialization of the newborn to the herd. However, exposure to so many animals only makes getting the cria on a vaccine schedule that much more pressing. Have your veterinarian look over the dam (female that just gave birth) to give her a clean bill of health. After two weeks of recovery, she will be ready to breed again.

Prompting Sexual Behavior
There are two methods of breeding. You can simply let your male alpaca out in the field with a harem in there as well. Nature will take its course. Because alpaca's are still relatively rare and expensive, very few farmers outside of South America will permit this. For a business enterprise, it is not the ideal way to breed desirable traits- the whole free choice thing usually does not line up with agricultural philosophy. For pets though, this is perfectly fine, as long as you do not have an image set in your mind for what the herd should look like after a few generations.

More commonly in the United States- males are kept separate from the females until the owner is ready to breed them. Then, they simply stick the desired male and female in an enclosed pad and let time take its course. Eventually, the female alpaca will become impregnated. This is usually more for controlled breeding programs than for just wanting to make another pet. Also remember, *Camelid* genetics are complicated, so you will most likely not get the cria you had in mind after any particular coupling.

Alpaca's are socially gregarious, and therefore, horny creatures. You will not need to do anything special to encourage them to breed. Keep them in good health, provide clean food and water, adequate space and let them mature, and they will figure it out eventually.

Genetics
As stated earlier, *Camelid* genetics are complicated. Coloration is complex, much like in humans. There is a spectrum of possibilities,

and sometimes you will have a surprise appearance of an old color pattern. However, there is one trait that is somewhat predictable when it comes to a breeding pattern- what kind of fleece they will produce (though, again, not necessarily its coloration). Suri fleece is dominant over Huacaya fleece (meaning this fleece type is recessive). Therefore, to get Huacaya fleece on your alpacas 100% of the time, you must cross breed two homozygous recessive Huacaya together. A simple Punnet square will let you know how to breed your alpacas, but you should still consult your veterinarian if you have your heart set on a particular type of fleece, as well as the exact pedigree of your alpacas when you purchase them. There are services that will do genetic testing on your alpacas to determine if they carry the Suri or Huacaya allele, and in what ratio- but be forewarned, it can be quite pricey!

Young Alpacas

Young alpacas are called crias. The come into this world after gestating for 11 months (even longer than a human baby!). They arrive into this world weighing only 12 to 14 pounds and are ready to go on their feet within an hour. They are almost always born one at a time, twins of any variety are exceedingly rare (far more so than in humans, if you wish to compare the frequency of double births on a per capita basis). Oddly, they are born on an ancient schedule, almost always during the middle of the day during the height of daylight hours. It is hypothesized this allowed their ancestors to dry off before the cool mountain nights came in. Regardless, care for your cria begins before it is even born.

Prenatal care
The dam (expected mother) should be vaccinated with the Clostridium CD/T vaccine prior to pregnancy. If this is not possible, the rule of thumb is to administer no vaccines in the 30 days running up to the expected due date of the cria. Also, if vaccines do have to be administered, it should be done so over a period of time. You have to remember that vaccines are designed to provoke an immune reaction, and too much going on in the immune system may be bad for the developing cria (possibly causing birth defects or miscarriage.).

Deworming of the dam is strongly suggested. Your baby cria will need all the nutrients it can muster during gestation, and parasites simply siphon these off for themselves. Discuss the deworming schedule with your veterinarian to figure out what schedule is best and what parasites you should worry about in your region. Well in advance of the expected due date, please trim your dams nails. Sharp nails can very easily injure a newborn cria (clearly unintentionally, but no one is awarding alpacas prizes for their foresight). Finally, make sure your alpaca has an ultrasound performed on it at least four times during the course of the pregnancy to ensure that the cria is developing appropriately.

Birth day!
Make sure you have your birthing kit ready. Momma may need some help from you, and you do not want to be caught off guard when it is time for the cria to enter the world.

- Scissors
- Heavy duty thread – for tying off the umbilical cord
- Rectal thermometer- To provide vital data
- Water-based lubricant (like K-Y jelly) – don't use a petroleum-based lubricant like Vaseline.
- Hair dryer- Air drying is fine in the Andes Mountains, but who knows what the weather is like where you are
- Iodine or isopropyl alcohol – Things need to be cleaned
- Spray bottle to spray iodine or isopropyl alcohol solution where needed
- Blanket – should be self-explanatory
- Towels – Again, self-explanatory
- Plastic feeding bottle and appropriate nipples. The cria will be hungry.
- Scale- You want to know how big baby is, right?
- Frozen colostrum- If nursing is delayed for whatever reason, it is important to have a meal ready to eat for your cria.

If it is your first cria, have the veterinarian there with you. It can be an overwhelming experience and is a legitimate medical emergency. Even if you have experience at birthing other animals, should you see your alpaca in the extreme discomfort of pain (like biting at the abdomen, straining excessively, rolling over completely, etc.) call your veterinarian immediately- something is wrong. Once labor has commenced, you can expect your cria to arrive in 30 to 45 minutes.

Once the baby cria has been delivered, be sure to dry it off as quickly as possible and put it under a heat source (like a heat lamp). Wait 20 minutes, then use your trusty thermometer to measure how warm your cria is internally (rectally). If your cria's body temperature is not yet at 99.5°F (35 C), then you need to warm up your cria- quickly! Fill a bathtub with water that is between 103-105°F (36.5-37.5 C), place your cria inside a large construction quality black garbage bag (WITH ITS HEAD OUT OF COURSE) and submerge the crias body in the water. The bag is designed to keep it warm, but the water should bring up the body temperature. Wait 10 to 15 minutes and recheck the body temperature. Once it is 99.5°F (35 C), your cria is in the clear. If the temperature is 68°F (20 C) or above, the cria and dam can go outside.

You should also squeeze the dam's teet. If no colostrum is being produced, you need to call the veterinarian. Make sure the dam has passed the placenta- should she fail to do so within 8 hours, you should call the veterinarian. Usually, the placenta will pass within 4 hours, so after that, you should start to worry. Also, if the placenta does not come out of its own accord- DO NOT YANK IT. This may cause a fatal rupture somewhere in your dams reproductive tract. If for whatever reason the cria cannot get colostrum from its mother, use frozen colostrum (thaw it out and bottle feed it- yes goat or cow colostrum is fine). If you cannot allocate colostrum for some reason, you have to get a hold of llama plasma and feed that to your cria (despite being rarer than cow colostrum, for sure).

Your crias first week of life should also follow a certain pattern. The vaccination should include Clostridium C, D, and Tetanus. Also, speak with your veterinarian about any other appropriate vaccinations and diseases of concern in your particular local area.

Rejection of Crias
Luckily, rejection is rare. However, it does happen. Many species of livestock engage in this behavior, and no one is quite sure what prompts a mother to simply walk away from offspring. Although a

heartbreaking scene to witness, if the cria has been abandoned, it is your responsibility as the owner to step up and take care of the unfortunate animal. You will need to feed it colostrum for as long as the veterinarian suggests to do so, and nurse it thereafter with an alpaca appropriate formula. It won't be long days. Fortunately, just another routine to add on top of your regular feeding regimen.

What to expect
Luckily, once you get past the first week with your cria, it can basically be treated as any other alpaca (except for breeding purposes, obviously). They are very sweet natured animals, and they grow up quickly. Female alpacas are fully-grown by 10 months, with the males usually fully maturing at about 2.5 years in. You need to handle your crias frequently to establish a trusting, loving relationship early on. If you fail to do this, doing anything literally with an older alpaca becomes much harder than it had to be. Chemistry takes time to develop between owners and pets, and alpacas are no exception to this rule.

Old Alpacas

It is a sad fact of life- everyone and everything gets older. Usually faster than we want to. Luckily, as far as pets go, alpacas are relatively long-lived, usually living between 15 and 20 years. Luckily for you as a pet owner, alpacas are also unique amongst livestock. Most livestock industries will slaughter animals well before they reach an elderly age- as soon as they cease to become productive for a secondary product (milk, eggs, wool), most are shipped off immediately to the slaughterhouse. It is true, there is a niche alpaca meat market, to even call it niche is overblowing how popular it is. It is extremely tiny. So, the alpaca fleece industry actually has a large body of literature and experience on care for elderly alpacas, as frankly, they do not know what else to do with them. Some of the more heartless businesses will simply give them away, ripping the alpaca away from the only herd it has ever known, luckily, most are simply put out to pasture and eventually euthanized when life becomes too much to bear for a senile alpaca.

The most common non-euthanasia cause of death amongst elderly alpacas is immune senescence. Basically, they get old and like almost every other bodily function, the immune system becomes weaker. Minor slip ups on either the alpaca's part or the owner's part become a major crises. Unfortunately, there is no way to fight it, it is the universe's entropy working against your pet alpaca, all your other pets and you as well. All you can do is look for changes in your pet alpaca's behavior and adjust accordingly. When the time comes for death, assuming you do not just simply find your pet keeled over in its stall or out in the field, call your veterinarian and do a proper euthanasia. It is the least you owe your pet.

Changes in eating and drinking

Alpaca's do a lot of chewing. It cannot be emphasized enough how much chewing they must do. They are herbivores, and most plant material (especially traditional forage like grass and hay) frankly has a much lower nutrient density than foods found in more omnivorous diets do. Years and years of eating tough foods and moving the jaws in a figure eight pattern to properly process said food will leave its impression on your pet alpaca. The front teeth will grow more slowly, and the chewing molars will become flatter from wear and tear. To offset the drop in the volume of food, you will need to provide a richer more nutritious feed so that your pet does not suffer from malnutrition.

Besides better nutrition, you can also separate your elderly alpaca, so it does not face competition for choice of food items. The young alpaca's have no tact and no sense of respect for their elderly, they will simply muscle in and take what they view as a treat. Stay on top of parasite control as well, what is a nuisance or uncomfortable to young alpacas can be downright fatal in elderly alpacas. Probiotics are also a key tool in aiding your elderly alpaca in getting the most possible out of its food. You may also want to supplement with additional electrolytes to make your elderly alpaca drink more- just as in humans, when they become elderly, sometimes they do not realize how dehydrated they are.

Changes in your Alpaca's appearance

The fleece will lose some of its luster, and it will grow more slowly. From a distance, you will not be able to readily tell an old alpaca apart from appearance alone. You will have to watch the body language, just as in humans, everything will start slowing down.

Unfortunately, arthritis is common in older alpacas- there is not a whole lot you can do to remedy the situation, only alleviate it. Feed it Cosequin and anti-inflammatories, discuss the regimen and frequency with your veterinarian to make your alpaca's quality of life. If you notice a change in the animal's weight, put it on the special feeding regimen above. Its appetite will decrease, this is perfectly natural. If weight loss is persistent despite earnest attempts to alter your alpaca's diet- you may want to call the vet in for a closer examination. Weight loss is symptomatic of many things, from minor parasitic infections to advanced cancer. It is best to get a frank opinion with which to make a better decision. If your vet recommends euthanasia, it is probably for the best. Never hesitate to get a second opinion though, if you have any persistent doubts. This is a simple kindness you can afford.

Chapter 12: Diseases that can afflict your Alpaca

Most livestock ailments can affect your alpaca. Luckily, most livestock illnesses have treatments! Treatments are described in detail earlier in the book, you should review them. Below is a sort of greatest (or worst, depending on your perspective) hits list of illnesses that can impact your alpaca's health.

Common Ailments

Congenital Defects

- **Choanal atresia-** This is an internal defect, it is when the internal nares do not form appropriately and make it difficult for your alpaca to separate eating and breathing chambers in its face respectively. Surgery is usually not suggested.

- **Wry face-** Basically, a misaligned face. You will know it when you see it. If severe enough, euthanasia is usually suggested.

- **Juvenile cataracts-** Cataracts will form from genetic causes, and impair vision. Discuss with your veterinarian the best course of action to fix this.

- **Short ears-** This is a dominant trait, very heritable. No corrective action is needed, it is also colloquially called "gopher ears."

- **Ventricular Septal Defect-** A hole in the wall separating the two lower chambers of the heart. Depending on severity, it may be treatable.

- **Syndactyl-** Fused digits. Depending on severity, the veterinarian may either recommend no course of corrective action or a minor surgery to split the fused digits.

- **Polydactyly-** Extra digits! Again, depending on severity, may be removed or left on.

- **Unilateral absence of a kidney-** Missing one kidney. Usually not a huge deal, but an interesting biological quirk.

Bacterial Diseases

- **Brucellosis-** Highly contagious bacterial infection. Luckily, it is somewhat rare, but only because of dedicated control efforts. This may spread from your alpaca to you! There is a vaccine- for your safety and your family's safety, please have your alpaca's vaccinated if your area is known to have endemic brucellosis.

- **Tuberculosis-** A very serious respiratory infection. Life-threatening to humans as well. Needs immediate quarantine and treatment. Animals may be destroyed to contain its spread. Luckily, due to extremely stringent animal movement controls, it is exceedingly rare in developed countries.

- **Anthrax-** Spore forming bacterial infection spread by animal or insect bites. Exceedingly rare due to extreme control measures- it is highly dangerous as well. It is a bioweapon of choice for premier military powers for a reason. If endemic where you are, please vaccinate.

- **Streptococcus zooepidemicus-** Respiratory infection far more common in South America than elsewhere- it is known as alpaca fever. Usually, stress weakens the immune system of an alpaca, and they catch a fever.

- **Tetanus-** Bacterial infection of the nervous system, causes severe muscle pains as well. Not contagious, but the bacteria is pretty ubiquitous in the environment, and thus, easy to catch from small open wounds. A very simple vaccine to administer (and thus prevent the disease), a very hard infection to treat after the fact.

Viral Diseases

- **Equine herpesvirus 1-** Associated with cohabitation with horses. May cause neurologic issues, and occasionally, blindness.

- **Bluetongue Virus-** Insect-borne disease, usually spread by midges. Symptoms of infection include high fever, excessive salivation, and swelling of the face and tongue. It is the excessive swelling that causes the tongue to turn blue, unfortunately. Incubation time is 5 to 20 days, and mortality is highly dependent on the individual animal's robustness. Recovery is usually slow. Vaccines are available for certain strains.

- **West Nile Virus-** Spread through mosquito bites, this virus impacts. Just as in humans, the majority of infections are asymptomatic. If your alpaca is not asymptomatic, however, the symptoms are very severe. Usually, it causes inflammation of the meninges (the layers of material surrounding the brain) and can cause death. Horse vaccines have been found to be equally effective in alpacas in safeguarding them against this virus.

Mycoplasma and Fungal Diseases

- **Mycoplasma haemolamae-** Causes anemia. Spread through dirty needles and insects. Usually, tetracycline administration helps manage this infection.

- **Coccidioidomycosis-** Known colloquially as "valley fever," this fungus is endemic to the American southwest, and can occasionally be found elsewhere. Symptoms are general-fatigue, cough, fever, etc. It may cause long-lasting lung problems if not treated quickly enough.

Gastrointestinal Issues

- **Megaesophagus-** Symptoms include weight loss and persistent regurgitation. No rhyme or reason as to why it happens, and no persistently successful course of action to remedy it. Come up with a custom game plan with your veterinarian and monitor from there.

- **Ulcer-** Usually diagnosed postmortem. Symptoms include weight loss. No coherent cause has been pinned down but usually associated with a severely stressed animal.

- **Stomach Atony-** Unknown cause. Symptoms include weight loss, fasting, and depression. Usually, the veterinarian will apply some weapons-grade probiotics through gavage into the rumen of the alpaca directly, and this treats it nicely. Regular probiotics will not cut it.

- **Hepatic disease-** *Fasciola hepatica* AKA the sheep liver fluke is your alpaca's liver worst enemy. The little critter will shut down your alpaca's liver. Unfortunately, due to the rapid evolution of this animal, there appears to be no vaccine. Luckily, it is exceedingly rare.

Skin Issues

- **Shearing injury-** Be more careful when shearing.

- **Sunburn-** Usually occurs after shearing. It will usually blow over on its own, but feel free to apply aloe vera if you are feeling sympathetic.

- **Ulcerative Pododermatitis-** AKA, "immersion foot." Symptoms include footpad blistering and sloughing. Basically a really nasty bacterial infection of the foot- usually treated with penicillin.

- **Skin parasites-** Ticks, mange mites, lice, they all love your alpaca's fleece as much as you do. There is a world of anti-parasitic over the counter treatments. Unless you are extremely neglectful, these infections are rarely fatal and easily treatable. They just spread like wildfire through a herd and are a big chore to solve, especially if you let it go too far.

- **Copper deficiency-** You can probably guess as to how to solve this one. Symptoms include poor growth in juveniles, depigmentation of fleece, and wiry texture to the fleece as well. Be careful with supplementation, as too much can lead to toxicity. Consult with your veterinarian.

- **Ulcerative Pododermatitis-** Lesions with crust. They can appear at any age, and no one is quite sure what causes them. Treatment is applying 1 gram of zinc sulfate a day or 2 to 4 grams of zinc methionine. Alfalfa hay should be discontinued and calcium supplementation reduced.

- **Munge-** Usually occurs in alpacas between 6 months of age to 2 years. Basically, an ugly muzzle, characterized by heavy crusting with lesions periodically. Treatment consists of taking care of secondary infections (as primary causes can be a range of things). Discuss with your veterinarian about the best course of action, as this is a highly specific disease.

Psychological Ailments

- **Depression-** Usually the result of trauma or digestive issues. Changes in herd structure too can cause depression. Consult with your veterinarian as to the best course of action for treating your specific alpaca's depressive symptoms.

- **Stress-** What on Earth can an alpaca be stressed about? Well- fighting, poor diet, perceived threat of predation. A lot. Keep a sharp eye out for what your alpacas are responding to negatively in their environment. If you ignore the causes of stress on your alpaca for too long, they will become susceptible to more severe ailments.

Chapter 13: Internet Resources

The Internet is a tremendous resource for your alpaca. It is a great way to find breeders, veterinarians, and supplies. Always take with a grain of salt what you read out in the World Wide Web, however.

Your Veterinarian should be your first source

Regardless of what you find online, your veterinarian has literally built a business on improving animal's conditions. He has your alpaca's best interest at heart. Even well-meaning accurate advice from the Internet can be partially informed, as very often there are geographic components to your alpaca's wellbeing. Your veterinarian is almost certainly a local who can provide yet another dimension to the care your alpaca receives. He or she may know what diseases are endemic, and which ones have lottery-like odds of affecting your herd. Your veterinarian is also a tremendous resource for learning proper husbandry skills- you should have your veterinarian at your side for every major "first" with your alpaca herd (first births, first delousing, etc.). If you are ever in doubt, call your veterinarian and run your thoughts by him or her. If in doubt of your veterinarian, get a second opinion. These are licensed reputable professionals who have built a career and lifestyle out of improving animal's conditions- you are in excellent hands.

Alpaca organizations you should know about

There are many organizations throughout the world dedicated to spreading knowledge and proper husbandry techniques for alpacas. Frankly, alpacas are only getting more popular, so more and more resources are popping up dedicated to proper alpaca care.

The United States and Canada
North America is home to the second largest population of alpacas in the world, only the native home range of South America has more. Although introduced relatively recently (in the 1980's and 1990's for the most part), they have caught on like wildfire. Unfortunately, not all owners are scrupulous, and there have been abandonment and cruelty issues. People very often get involved with animals without

understanding the true depth of commitment it is to be a proper pet owner. So please, if you are thinking of acquiring alpacas, whether it be your first or your hundredth, consider adopting. There seem to be rescue organizations in just about every state and province- do a local search. Below we will name some of the more prominent institutions.

Rescue Organizations

- **Cross Creek Alpaca Rescue-** Founded in 2009, it has rapidly become the premier alpaca rescue in the United States. Found in Tenino, Washington they are a fully tax exempt organization, meaning donations are a wonderful way to claim some tax relief for yourself come April! Their website is www.crosscreekalpacarescue.org

- **Beaver Creek Farm Sanctuary-** Based out of Wainfleet, Ontario this is a rescue devoted to many species. They do seem to be the premier alpaca rescue in Canada at the moment. Their website is beavercreekfarm.com

Informational Organizations

Alpaca Owners Association- This is the one stop shop for connections in the alpaca world. They have an extensive registry of alpaca farms and rescues from which to draw alpaca related wisdom from. They maintain genetic databases, promote alpaca related research, market alpaca products, you name it, they do it. Headquartered in Lincoln, Nebraska, and the organization can be traced back to predecessor organizations in 1987 and 1991 respectively, the AOA is a not-for-profit corporation with 10 employees and a member board of directors. Extremely professional, they are a tremendous wealth of information on everything alpaca related. Their website is alpacainfo.com.

United Kingdom
Alpacas are relative newcomers to the United Kingdom, but luckily the moderate climate has made an ideal home away from home for alpacas. To be honest, it was the United Kingdom who has arguably

done more for alpacas than any other nation in promoting the fleece in western markets, all the way back to the 19th century. Nowadays, it appears to be quite the growth industry!

Rescue Organizations

- **Little Hamlet Alpacas Rescue Centre-** Based out of Cornwall, this appears to be the premier alpaca rescue organization in the United Kingdom. Founded in 2004, they have a substantial herd of rescue animals. Also, they offer educational opportunities, with half-day training courses running £50 per couple- an excellent investment if you intend to own alpacas. You also have an opportunity to stay with them if you intend to adopt from them- they seem to be quite serious in their dedication to imparting proper husbandry techniques. This is their website: http://littlehamletalpacas.co.uk.

Informational Organizations

- **The British Alpaca Society-** Representing 1400 members who in turn own 35,000 alpacas cumulatively- there is no arguing this is the premier British Alpaca Society! They are responsible for industry publications and promoting the alpaca industry both domestically and throughout Europe. They host shows to showcase alpacas, what's not to love? This is their website: http://www.bas-uk.com/.

Australia and New Zealand

- **Australian Alpaca Society-** it is all in the name. Based out of Melbourne, Victoria, they promote all things alpaca within Australia. They organize shows and encourage their members to share advice, they can point you in the right direction of whatever answers you seek. This is their website: https://www.alpaca.asn.au/.

- **Alpaca Association New Zealand-** The chief organizers for everything alpaca in New Zealand. This is their website: http://www.alpaca.org.nz/